Crossing the Threshold

Advancing into Space to Benefit the Earth

Paul O. Wieland, P.E.

Threshold 2020 Press

All rights reserved.
Published in the United States by
Threshold 2020 Press
Huntsville, Alabama
www.threshold2020.com

Library of Congress Cataloging-in-Publication Data
Wieland, Paul O.
   Crossing the threshold: advancing into space to benefit the earth/
Paul O. Wieland.—1st ed.
   Includes bibliographic references and index.
   ISBN: 978-0-9825127-0-8 (hard cover)
   ISBN: 978-0-9825127-1-5 (paperback)
   1. Space exploration. 2. Environment. 3. Technology. 4. Energy.
5. Resources. 6. Current Affairs – 21st century.

Library of Congress Control Number: 2010903938

Printed in the United States of America
on paper containing recycled content.

Cover images courtesy of NASA (space elevator and Earthrise).

The cover depicts a concept of the space elevator (in the top panel) in transit from the surface of the Earth to geosynchronous orbit. The center panel shows a view of the sky through a geodesic dome. The lower panel is the Earthrise photo taken by the *Apollo 8* crew during their orbit of the moon.

To those who have given their lives
advancing our efforts to explore and live in space,
and to all who strive for a sustainable and just world.

Also by Paul O. Wieland:

Designing For Human Presence in Space
An Introduction to
Environmental Control and Life Support Systems
NASA RP-1324, 1994

Living Together in Space
The Design and Operation of the
Life Support Systems on the International Space Station
NASA TM-1998-206956, 1998

Living Together in Space
The International Space Station
Internal Active Thermal Control System
Issues and Solutions—
Sustaining Engineering Activities
at the Marshall Space Flight Center, 1998 to 2005
with M.C. Roman and L. Miller,
NASA TM-2007-214964, 2007

These reports, and others by Paul Wieland, are available through the NASA
Technical Report Server, http://ntrs.nasa.gov/search.jsp

Acknowledgments

Many people contributed directly and indirectly to this book. I can never adequately acknowledge their contributions. Special thanks for their suggestions, comments, and encouragement to Paul Johnson, Don Ford, Cynthia Robinson, Jennifer Madden, Bob Goss, Tom Guffin, Linda Haynes, Jay Perry, Perry Cartwright, Katherine Fausset, Marda Burton, Lori Jones, Rebecca Smith, Stacey Haire, Rebecca Zurn, Randy Humphries, James Perkins, Jim Griffith, and Alice Mello. I deeply appreciate their interest and support. Editing was performed by Molly Felder and the cover was designed by Michael Wieland.

About the Author

Paul Wieland is a professional engineer (P.E.) who earned degrees in botany and mechanical engineering from the University of Louisville (Kentucky). From 1983 to 2005, he was a NASA civil servant at the Marshall Space Flight Center in Alabama, working primarily on development of the environmental control and life support systems for the International Space Station and other space missions. In 1991 he was a founding member of the Institute for Advanced Studies in Life Support. Pursuing his interests in sustainability and energy efficiency, from 1989 to 1991 he founded and ran Wiseland Products, the first paper company in Alabama to carry recycled paper products; he served as a technical adviser to the American Lung Association Health House® '96 project in Huntsville, AL, to demonstrate that it is possible to build a house that is energy efficient and has good indoor air quality, and later renovated three 60-year-old houses to meet the Health House criteria; and, in 2009, he became a LEED Accredited Professional. He also converted a Ford F-250 diesel pickup truck to run on waste vegetable oil. Paul enjoys hiking and camping; contra dancing and swing dancing—he founded the Huntsville Swing Dance Society in 1995; and making pottery—he co-founded the Tennessee Valley Ceramic Arts Guild in 2008.

Contents

Preface / ix

Introduction: The Future Begins Now / 1

1 A World of Abundance / 4

Part I: Thresholds / 11

2 Personal Thresholds / 13
3 The Lessons of History / 22
    Geographic Thresholds / 23
    Technological Thresholds / 28
    Thresholds of Challenge / 34
    Thresholds and Change / 41
4 Risk: The Surly Bonds of Earth / 44

Part II: Envisioning the Future / 48

5 Visions for the 21st Century / 50
6 New Challenges / 58
7 New Ideas / 66
8 The Promise of Space / 79

Part III: The Way to the Future / 83

9 The Means, Part I: Getting to Space / 84
    The Limitations of Rockets / 85
    New Ideas for Reaching Space / 87
    Leaving Earth Orbit / 96
10 The Means, Part II: Living in Space / 102
11 The Means, Part III: Co-opetition / 112
12 Convergence / 120

Part IV: Creating the Future Now / 124

13 Leadership in Space / 126
14 Space Commerce / 130
15 Resources from Space / 138
16 Manufacturing in Space / 142
17 Power from Space / 149
18 Technology to Reach Space / 153
19 International Agreements / 156

Part V: Roles for All / 160

20 Government Roles: Vision and Policies / 162
21 Education and Academia: Discovery / 166
    Learning / 167
    Inspiring / 168
    Discovering / 172
22 Private Enterprise: Developing Products and Providing
    Services / 174
23 Your Role: Creativity and Action / 180

Conclusion: A New Beginning / 185

24 A New Age of Discovery / 187
25 Crossing the Threshold / 191

References and End Notes / 194

Index / 256

Preface

ARRIVING HOME FROM WORK ONE THURSDAY in mid-January 2000, I found a phone message—from the White House. A staffer had called on behalf of President Bill Clinton, who was asking leading authorities to describe the futures of their fields. The responses were to be compiled into a book for the president: *Visions of the Future from Leading Thinkers*. A "one- to two-page memo directly to the president" was requested on how I see space activities "coming along in the next 50 to 100 years or even further." Honored and excited, I composed my thoughts and sent the memo by the requested date, the following Monday.

Such a request from the president is not to be refused, but discussing a 50- to 100-year period is a formidable task and I was not satisfied with the result of my effort. Three days were simply insufficient time to do justice to the topic, and two pages to discuss our future in space provided too little, well—space—to do so properly. Only in a book could I adequately respond. This is that book.

Many books have been written about possible futures in space and the benefits to be obtained through space activities. Some are technically intriguing, while others are pure fantasy. Even fantasy, however, can be valuable by stimulating the imagination, and that may be the primary goal.

THIS BOOK PRESENTS A VISION of a world of abundance—of energy, resources, and opportunities—with space activities contributing significantly. More importantly, the primary goals are to identify the critical factors and examine the decisions that must be made and the actions that must be taken to enable that future. In the not-too-distant future, we could have the capability to begin utilizing space to directly address critical needs on Earth. How we choose to develop that capability will determine how successfully we deal with the challenges before us.

A vision that addresses those challenges is the first step toward successfully resolving them. Knowledge and understanding are also essential, along with inspired curiosity, creativity, ingenuity, and the desire to reach for the vision.

Whether you are interested in our future on Earth or our future in space, whether you are in school, pursuing your career, or a member of Congress, my hope is that this book will stimulate discussion, elicit fresh ideas, and inspire you to action. A new Age of Discovery, a new Renaissance, is within our reach, by proceeding into space with the goal of addressing the issues of the 21st century in ways that enhance opportunities for everyone on Earth. In doing this, we will ensure a brighter future for all.

Paul O. Wieland

Introduction

The Future Begins Now

"In my own view, the important achievement of Apollo was a demonstration that humanity is not forever chained to this planet, and our visions go rather further than that, and our opportunities are unlimited." — Neil Armstrong, astronaut, July 1999

AS WE ENTER THE SECOND DECADE of the 21st century, we are at the dawn of new opportunities to address age-old challenges, as well as new ones, and to become a space-faring civilization in the process. The decisions we make about how to proceed will determine the characteristics of the future that unfolds. Cooperation will be essential due to the scope of the challenges. Providing opportunities for broad involvement and ensuring that benefits are broadly distributed will improve our chances of successfully meeting the challenges.

• • •

SEPTEMBER 11, 2001 MAY BE THE "REAL" DEMARCATION of the 21st century. The terrorists who hijacked four airplanes and flew three of them into the World Trade Center in New York City and the Pentagon in Washington, D.C., provided a rude awakening to the realities of the new century. The Cold War and the Soviet Union are history, over for almost 20 years now, but conflict has not ended. The possibility of a major conflagration between superpowers has waned, but other conflicts that had been in the wings have taken center stage. Addressing the challenges of the 21st century involves dealing effectively with these conflicts.

The methods used in the past may not be so effective now. Brute force, while capable of subduing, has its limits, a lesson that we should learn from multiple wars (with multiple aggressors) in the 20th century. In a globalized world, conflict in one locale has broader impacts, especially with the threat of nuclear weapon proliferation. Disputes over energy, resources, and environmental changes will tend to exacerbate other conflicts, especially when water and food supplies are adversely affected. As the developing world demands more energy and resources to support their entry to the developed world, societal pressures will also increase. By ensuring abundant energy and resources, without competing for essential commodities or degrading the environment, we can mitigate the conflicts. By promoting ways for all to live in dignity we can turn conflicts into opportunities.[1]

The challenges before us are great, but I am optimistic about our future. I believe that we can effectively address the challenges. I believe, too, that becoming a space-faring civilization is an integral part of achieving the best possible future. As David Livingston, host of *The Space Show* radio program, says, "India, Russia, China, Japan, and the European Space Agency ... all want a manned mission to the Moon and it won't stop there. These countries and agencies know that manned space exploration builds wealth for their nation, solves problems and enhances life for their people right here on Earth, and shows us the way for how we can all live together in peace... It's not just about what we learn out there in space, or about ourselves, or how to be a better steward of precious Earth. It's about how we live here on Earth together and what type of future we want for ourselves and our children." We have the ability to create a sustainable world with virtually unlimited energy and resources, and with unprecedented opportunity for all. To do so, thoughtful action is needed.[2]

DUE TO THE GROWING URGENCY OF THE CHALLENGES we face, the next 10 years are especially important, and that is the timeframe of the actions proposed in this book. Our choices now will be critical in determining our future options, and will depend on how we answer important questions, including:

- What are the fundamental issues of the 21st century?

- How cooperative or competitive—between countries and between governments and private enterprise—should our efforts be when dealing with those issues?
- What technological advances could be beneficial for addressing those issues?
- Which issues could our efforts in space directly address?
- What roles should governments, private enterprise, and academia play regarding space activities?
- How should the risks and benefits of space activities be distributed?

The answers will be chosen by our leaders, but we—individually or in groups—can affect their choices by letting our views be known. What we decide as individuals makes a difference and, so, in a very real sense each of us determines the future.

THIS BOOK IS DIVIDED INTO SEVERAL PARTS: Part I provides background that supports the later proposals, though it may seem to be a digression from the main theme of utilizing space to address societal issues. (Later chapters are still readable if it is skipped.) Part II presents a vision of a sustainable future and conveys how space activities can help to attain that vision. Part III addresses the means by which we may achieve the vision for space activities and describes current efforts to develop those means. In Part IV specific recommendations are made to promote utilization of space to address key issues. And Part V describes the roles of governments, private enterprise, academia and researchers, and individuals (you!) and groups, to achieve the vision. Personal information—including descriptions of my experiences while working for the National Aeronautics and Space Administration (NASA)—is presented where related to particular points being made.

How we imagine our future will affect the decisions that we make and the results that we obtain. Let's imagine a world of abundance and find ways to make it happen.

I

A World of Abundance

"[T]echnologically humanity now has the opportunity, for the first time in its history, to operate our planet in such a manner as to support and accommodate all humanity at a substantially more advanced standard of living than any humans have ever experienced." — Buckminster Fuller, *Grunch of Giants*, 1983

IMAGINE THE WORLD WITH ABUNDANT, AFFORDABLE ENERGY that does not contribute to environmental degradation, hazardous waste, or security issues. Imagine the world with abundant resources that are acquired in ways that do not damage the environment. Imagine the world where the air and water are clean, forests are verdant wildlife habitat, the oceans are healthy with abundant fish populations, food is plentiful and nutritious, farmland is sustainable, and cities are pleasant places to live. Imagine the world better for the generations that follow us, with abundant opportunities for all to live more fulfilling lives.

• • •

IN 1956, M. KING HUBBERT, a geophysicist at Shell Oil Company, forecast that United States domestic oil production would peak between 1966 and 1970. He was scoffed at for his prediction, but he was vindicated when U.S. oil production did, indeed, peak in 1970, leading to steadily increasing imports of oil. His prediction may be considered the seminal event regarding fossil-fuel limitations. Published in 1962, Rachel Carson's *Silent Spring* describes the devastating consequences of pesticide use on wildlife and may be considered the seminal event for modern environmental issues related to the effects of our activities. Convergence of these two events is evident with the 2007 release of the Intergovernmental Panel on Climate Change (IPCC) report on global warming, concluding that human activities are a significant contributor, especially the burning of fossil-fuels which releases carbon dioxide (CO_2) into the atmosphere. (Other gases such as methane and soot from open fires also contribute to global warming.) The acknowledgment that human activities are causing major changes in the Earth's environment is a recognition of our ability to transform the Earth, and this is one of the fundamental issues of the 21st century. By the choices we make, we determine whether the transformation degrades or improves the natural environment and our quality of life.[3]

Ultimately, the goal of humanity is to ensure our long-term survival and that requires preserving enough of the Earth's natural ecosystems to keep the Earth viable. Given our current state of ignorance of the complex interconnections between the various parts of the Earth's ecosphere, what are we risking by continuing to burn fossil fuels, waste resources, cut down or burn old growth forests, overfish the oceans, and contaminate the air and water and even the food that we eat? The toxic waste products that we release into the environment may last for thousands of years. Imagine the ancient Greeks or Chinese leaving waste that would still be hazardous to us today. "Business as usual" cannot continue, but the future need not be one of want and limitation.

SUSTAINABILITY WAS SUCCINCTLY DEFINED by the United Nations in 1987 as "meeting the needs of the present without compromising the ability of future generations to meet their own needs." Recent efforts point the way to a sustainable future of abundant energy and resources, with minimal environmental impact. Actions by the U.S. (and other governments) include efforts to improve energy efficiency, such as by raising the automobile fuel efficiency standards, by instituting efficiency standards for appliances and buildings (e.g., the EnergyStar program of the U.S. Department of Energy (DOE)), and by funding research to develop more efficient lighting (e.g., light emitting diode (LED) lamps). The Kyoto Protocol, the Copenhagen Climate Change Summit, and other efforts aim to reduce generation of CO_2 by providing incentives for using renewable energy sources and for implementing conservation measures.[i] Private organizations are also promoting energy and resource efficiency, such as the U.S. Green Building Council through their Leadership in Energy and Environmental Design (LEED) rating system for buildings. Similarly, other recent efforts point the way to a future of resource abundance by recycling materials, by designing products for reuse at the end of their life ("cradle-to-cradle"), and by using materials more efficiently.[4]

Efforts to be "carbon neutral" are gaining support, by designing buildings or entire cities to use so little energy that available renewable methods of power generation can provide all the power needed, or by removing from the atmosphere—generally by planting trees—the CO_2 produced during the manufacture of products or when performing other activities. Even in regions that currently export petroleum, some leaders know that one day their bonanza will come to an end and they are preparing for a post-petroleum economy by increasing energy efficiency, reducing fossil-fuel energy use, and recycling waste materials.[5]

In February 2008, the United Arab Emirates began construction of a new city on six square kilometers (2.3 square miles) of land outside of Abu Dhabi. Masdar City is designed to be a "zero-carbon, zero-waste, and car-free city" for 1,500 businesses and 50,000 residents. Power will be supplied from solar, wind, and geothermal sources; a solar-powered desalination plant will provide fresh water; wastewater will be recycled, including for crop irrigation; and much of the food needed will be grown on-site. One goal is to have zero waste by recovering nutrients from biological waste, recycling plastics and metals, and producing power from other waste. The $22 billion project will include a new graduate university dedicated to renewable energy, the Masdar Institute of Science and Technology, which is expected to be completed and fully functioning by 2015.[6]

---

[i] Toward this end the European Union, a consortium of 27 nations, began phasing out incandescent light bulbs on September 1, 2009 by eliminating the manufacture or importation of bulbs rated at 100 watts or greater, to promote the use of more efficient lighting.

In the U.S., similar efforts are underway. On May 4, 2007, Greensburg, Kansas, was devastated by an F-5 tornado. Eleven people were killed and 95 percent of the town was destroyed. Rather than move away or try to rebuild the town as it was, the residents decided to rebuild as a model of sustainability. As the old town was destroyed by wind, the new town will be powered by wind. With the assistance of the DOE Office of Energy Efficiency and Renewable Energy (EERE), the entire town will be powered by wind turbines providing 3 to 4 megaWatts (MW) of electricity, with backup generators using biodiesel fuel. To minimize energy consumption, all large buildings must achieve the highest LEED rating. The recovery plan for the town describes over 40 projects to revitalize Greensburg as a demonstration of sustainable living.[7]

Addressing energy and resource issues is vital, but not sufficient. Improving our quality of life, in general, is also important. Masdar City and Greensburg point the way to a sustainable future and show what is possible in specific situations to reduce energy and resource use while ensuring a high standard of living. To address the larger challenge, several groups have presented proposals for acquiring and using energy and resources on a broad scale in ways that minimize or eliminate environmental and other concerns. Many share the sentiment of Dolly Baus when she expressed her frustration, in July 2008, by saying, "We can put a man on the moon. Don't tell me we can't do a better job with energy."[8]

THE GLOBAL ENERGY NETWORK INSTITUTE PROMOTES the use of a range of renewable energy sources, from wind and solar to tidal flow and biomass. Proposing to acquire electrical power from these sources where they are abundant and interconnect the electrical grid to distribute the power where it is needed, the GENI position is that "There Is No Energy Crisis, There is a Crisis of Ignorance."[9]

In 2008, calling for a concerted effort reminiscent of the Apollo program to reach the Moon, former U.S. Vice-President Al Gore—whose movie *An Inconvenient Truth* called for action to address global warming and won him a Nobel Prize—claimed that within a decade it will be possible to stop using fossil fuels to generate electricity, by using solar, wind, and other renewable sources. Also in 2008, T. Boone Pickens, a Texan who made his fortune in the oil business, presented his somewhat less ambitious plan to use wind turbines to generate 22 percent of the electricity used in the U.S. This would allow the natural gas presently being used to generate electricity to be used instead for powering cars, buses, and trucks, which would reduce the amount of oil imported into the U.S. by 38 percent within ten years.[10]

Other proposals include those from Robert Socolow and Stephen W. Pacala with the Carbon Mitigation Initiative of Princeton University, who present 15 strategies, each of which could reduce global CO_2 emissions by one billion tons/yr; Amory Lovins of the Rocky Mountain Institute who promotes improving energy conservation, increasing efficiency, and using cellulosic biofuels; Lester Brown of the Earth Policy Institute who promotes improving efficiency and dramatically increasing wind generated power; Arjun Makhijani, a former adviser to the Tennessee Valley Authority (TVA), who claims we can eliminate fossil fueled and nuclear power plants and still meet our energy needs; and the G-8 International Partnership for Energy Efficiency Cooperation (IPEEC) formed in 2009 to support a market-based approach to reduce greenhouse gas emissions. These proposals (summarized in the end notes[11]) focus mainly on increasing efficiency and using wind, solar, and biomass energy, though some include using nuclear power and capturing and sequestering CO_2 from coal-fired power plants.

We need creative solutions, and that requires looking at the challenges in new ways. The International Energy Agency says that "technological breakthroughs that change profoundly the way we produce and consume energy will almost certainly be needed ... The sooner a start is made, the quicker a new generation of more efficient and low- or zero-carbon energy systems can be put in place." In addition to technological breakthroughs, incremental improvements can, cumulatively, produce significant benefits, and, therefore, are just as important. Breakthrough innovations and incremental advances could completely change the way that energy is supplied and used.[12]

The proliferation of proposals addressing energy, resources, and the environment indicates the seriousness of the situation. Implementing any of them will effectively increase the supply of energy and resources, and reduce our impact on the environment. But while there is no shortage of proposals, the slow pace of implementation indicates that there is a shortage of plans that *inspire* us to embark on the changes that are needed.

THIS BOOK IS ABOUT HOW SPACE ACTIVITIES could help address energy, resource, and other issues on Earth. A program of space activities is not *the* answer to the emerging problems of the 21st century, but it can be a significant and vital part of *an* answer, while providing benefits that other answers cannot. Space-related activities can provide new technologies, novel materials, and more efficient ways of performing tasks that can reduce energy and resource use on the Earth. These kinds of benefits occur now and have been occurring since the beginning of the space age. Not all of the space-related advances have been used to the extent that they could be, though, so greater efforts to commercialize existing space-related technologies would provide even greater benefits and would pave the way for rapid commercialization of new advances. However, spin-off benefits are not sufficient reasons for a space program.

Space activities are already critical to our efforts relating to global warming and resource usage. The satellites monitoring the condition of the Earth, the effects of solar activity, and the interaction between the Earth and the space environment provide information about changes in the Earth's condition, and spacecraft visiting other planets provide information about conditions on those planets that enables better understanding of the Earth's climate. But energy and resource issues also can be addressed more directly by activities in space. Above the Earth's atmosphere, radiation from the sun has several times the energy that reaches the ground. By capturing that energy in space and delivering it to the Earth, we could power society without degrading the environment. Mineral resources in space are abundant—on the Moon and in asteroids—and those resources could be mined, again, without adverse impacts to the Earth's environment.

In addition to the possibilities of utilizing space-based solar power and acquiring materials from space, activities in space can inspire us in ways that the proposals mentioned above do not, and broaden our perspective so we can see challenges in new ways and find new solutions. Space-related activities can also promote international cooperation that can support cooperation in other areas. Including space in our efforts to obtain energy and resources could also usher in a new Age of Discovery, as the development of ready access to space will enhance our capabilities for exploring the solar system at large. The challenges of utilizing space to address Earth-bound issues include developing an inexpensive, environmentally benign means of getting to space, and dealing with the challenges of being in space (radiation, micro-gravity effects, life support, etc.). By finding solutions to these challenges, we will cross thresholds that open up great opportunities.

Part I

Thresholds

"For forty-nine months between 1968 and 1972 two dozen Americans had the great good fortune to briefly visit the Moon. Half of us became the first emissaries from Earth to tread its dusty surface. We who did so were privileged to represent the hopes and dreams of all humanity. For mankind it was a giant leap, for a species that evolved from the stone age to create sophisticated rockets and spacecraft that made a Moon landing possible. For one crowning moment, we were creatures of the cosmic ocean, an epoch that a thousand years hence may be seen as the signature of our century." — Edwin "Buzz" Aldrin Jr., astronaut

• • •

IN 1956, WHEN ONLY A FEW MEN HAD FLOWN high enough to cross the boundary of space (the "threshold") and be awarded their astronaut wings—and no man-made object had yet orbited the Earth—a dramatic movie was released about our efforts to reach space. *On the Threshold of Space* took a behind-the-scenes look at the program to understand the physiological and psychological effects on people of moving at high speeds, ascending to extreme altitudes, and being confined in small volumes for durations representative of space missions. At the time, it was not known whether people could even live in space, so every possible measure was taken to ensure the best chance of survival. The success of those efforts was demonstrated by the *Apollo 11* landing on the Moon in 1969.[13]

There is another threshold of space that is not physical, but rather is a threshold of awareness. Those who have been in space have described the experience in various ways. Some have recalled that on the first day, when looking at the Earth, they looked for their country, on the second day they looked for their continent, but by the third day in orbit they saw the Earth as a whole. The common result is a transformed awareness and they realize that a forest fire or oil spill in one place ultimately affects the entire planet. From the ground, we see only a tiny portion of the Earth at a time and the entire Earth seems to be quite large. From space, however, when the entire Earth can be seen against the background of the vastness of space the scale shifts, producing what has been called the "overview effect." As James B. Irwin, an *Apollo 15* mission astronaut, described the experience: "As we got further and further away, [the Earth] diminished in size. Finally it shrank to the size of a marble, the most beautiful you can imagine. That beautiful, warm, living object looked so fragile, so delicate, that if you touched it with a finger it would crumble and fall apart. Seeing this has to change a man."[14]

Those of us not so privileged to directly view the Earth from space have a glimpse of that experience when we look at photographs of the Earth in space. As Aldrin says in the lead-in quote, we are "creatures of the cosmic ocean." Those who *feel* that truth, cross the threshold of awareness of our place in the cosmos.

THERE ARE OTHER THRESHOLDS AS WELL, crossing points, if you will, that once crossed can never be uncrossed. Whether in our personal lives or in society as a whole (which is the collective sum of the individuals in it) we may cross a threshold by exploring new territory, acquiring new knowledge, or creating something new, that changes our understanding and our options for further action. Crossing a threshold involves making a decision, selecting a particular course of action over the alternatives. Historical examples of crossing thresholds can provide insights relevant to utilizing space to address the challenges of the 21st century. Yet, even when a society-changing threshold is crossed, it begins as a personal decision by an individual. Such personal thresholds can have profound effects.

2

Personal Thresholds

"I cannot do everything, but I can do something. And I will not let what I cannot do interfere with what I can do." — Edward Everett Hale, author of *The Brick Moon* (1870)

"I long to accomplish a great and noble task, but it is my chief duty to accomplish small tasks as if they were great and noble." — Helen Keller

"Two roads diverged in a yellow wood and ... I took the one less traveled by and that has made all the difference." — Robert Frost, from *The Road Not Taken*

THRESHOLDS MAY BE CROSSED when we take some action or make a decision that changes our perspective on the world. As such, thresholds can relate to a rite of passage or other life-changing event, or to seemingly mundane decisions that have crucial results. Robert Frost's *The Road Not Taken* is a poetic expression of making a decision that may represent a personal threshold. Along with crossing a threshold, comes a realization of some sort—a new awareness, new abilities, or new opportunities.

• • •

IT WAS THROUGH A TWIST OF FATE that I came to work for NASA. Growing up in the 1960s, I shared the excitement of Walter Cronkite reporting on the space program, built model rockets, studied science, drank Tang® at breakfast like the astronauts,[ii] and

---

[ii] Tang®, an orange-flavored powdered drink mix, was invented by General Foods Corporation in 1957 and was selected by NASA for use in the Gemini program. Tang was mixed with water produced by fuel cells, which had a distinctive flavor, to make the water more palatable.

wanted to be an astronaut. Along with millions of other Americans, as well as millions of people around the world, I followed each stage of the effort to reach the Moon, thrilled as interim missions were accomplished and disappointed at the setbacks, which served to heighten the sense of urgency. The goal was dramatically achieved on July 20, 1969, just months before the end-of-the-decade deadline set by President Kennedy in 1961.

The 1960s was a decade of tumultuous change in the U.S. and in the world. The Cold War threat of the Union of Soviet Socialist Republics (USSR) was an underlying presence in everyday life, and the Vietnam War—a surrogate conflict for direct warfare with the USSR—spawned an anti-war movement. "Power to the people" became a catch-phrase for implementing social change; Martin Luther King, Jr. gave his "I Have a Dream" speech in Washington, D.C., presenting his vision of racial equality; and the Civil Rights Act was passed. Influenced by Betty Friedan's book, *The Feminine Mystique*, a resurgent women's movement sought greater equality for women. The assassinations of President Kennedy, in the midst of his term, and Martin Luther King, Jr., and Robert Kennedy (the brother of president Kennedy) in the prime of their lives, had profound effects on the country and the world. Also in the 1960s, "Beatlemania" became all the rage and the Woodstock music festival was a defining event; the National Environmental Policy Act was passed and the term "spaceship Earth" became part of our vocabulary; the television show *Star Trek* was an adventure drama set in the 23rd century that portrayed faster-than-light travel and beaming of objects and people, capabilities not supported by current theories of physics; and the movie *2001: A Space Odyssey* was released, showing a closer future that *is* feasible given known theories of physics. The world was changing quickly and it seemed that predictions of our steady advance into space could certainly be realized.[15]

In 1969, at the time of *Apollo 11* and the first steps on the Moon, I was 15 and intent on taking steps of my own, to make my mark on the world. By the time I graduated from high school in June 1972, however, it seemed unlikely that those steps would take me to NASA. The last Apollo mission to the Moon was only a few months away and NASA was downsizing, laying off engineers, scientists, and support personnel. Additional Apollo missions to the Moon were canceled, even after the Saturn V rockets to launch them had been built. We had reached the Moon, but the momentum stalled. NASA's bold plans for multiple space stations and other follow-up missions were scaled back to *Skylab*, the *Apollo-Soyuz Test Project*, and several ambitious robotic missions. Those missions, along with the upcoming development of the Space Shuttle, would require fewer personnel than the Apollo program. The decreasing workforce dimmed my hope of working for NASA.[16]

In the 1970s, the tense excitement of the previous decade ebbed and our attention shifted to different concerns. Spurred by events in the 1960s such as the publication of *Silent Spring*; the Cuyahoga River catching fire yet again; and the first picture of Earth rising over the horizon of the Moon—an iconic image taken during the *Apollo 8* circumlunar mission in December 1968 that sharply contrasts the colorful, life-sustaining Earth with the monochromatic, sterile lunar surface—awareness was increasing of our impacts on the ability of the Earth to support life. In 1970, the first Earth Day celebration was held and, in response to public concern over health issues, the Clean Air Act was expanded and the Environmental Protection Agency was created. In 1972, the Clean Water Act was enacted. Also in the 1970s, the U.S. ended military operations in Vietnam, the OPEC[iii] oil embargo occurred just as domestic oil production began declining (leading to gasoline shortages and higher prices), the Strategic Arms Limitation Treaty (SALT) established limits to the nuclear arsenals of the U.S. and USSR, and the Three Mile Island nuclear reactor in Pennsylvania leaked radioactive material into the atmosphere.[17]

BEGINNING IN THE FALL SEMESTER of 1972, I attended the University of Louisville, in Kentucky, but without a clear vision of

---

[iii] The Organization of the Petroleum Exporting Countries consists of Saudi Arabia, Iraq, Iran, Kuwait, Venezuela, Libya, Qatar, and several other countries. http://www.opec.org/library/faqs/aboutopec/q3.htm

where I was headed in life. In high school, I had enjoyed biology and had been in the Ecology Club, so I majored in botany, pursuing my interest in the natural world. But my interest was also with the thought that long-duration human spaceflight would require growing plants for air and water purification, as well as for food, and I hoped that the *Skylab* program would lead to a resurgence in human space activities. However, after my sophomore year—and after the last *Skylab* mission—I attended part-time and then took a break from school.

From 1978 to 1980, I worked at the East Louisville Recreation Center in the Clarksdale housing project, first for a crime prevention program (to involve the youth in activities to reduce crime, such as producing a neighborhood newsletter and cleaning up litter) and later for the Louisville Metro Parks Department. During that time, a group of engineers from the local General Electric plant formed a Clarksdale chapter of the Junior Engineering Technical Society (JETS). Though the residents of Clarksdale were poor, there was engineering interest and talent in the community. At an egg drop competition held in 1978 a middle-school student from Clarksdale won, with the only unbroken egg. The GE engineers heard about this and wanted to foster the interest in engineering. They met first at the recreation center to talk with the kids about the excitement of designing and building things and about the personal benefits of being an engineer. Their meetings moved to a nearby church and I never saw them again, but I began to think about the possibility of becoming an engineer myself.[18]

As a parks department employee, the city would pay my tuition, so I returned to school part-time, attending classes during the day and working in the evenings. As I was completing my degree in botany, I continued to think about pursuing engineering, and at the beginning of 1980 I enrolled full-time in the engineering school of the U of L, majoring in mechanical engineering. One of my first classes was Engineering Statics, about the forces relating to rigid bodies at rest or moving at a constant velocity. I had never taken a physics class before, so this would be a real test of my ability. I studied hard and on the first exam I made a perfect score. This was not as difficult as I had feared.[19]

IN A SHOW OF HOW THE SUN CAN AFFECT THE EARTH, on July 11, 1979 *Skylab* returned to Earth as a meteor over Australia. Greater-than-expected solar flare activity caused a slight expansion of the atmosphere, thereby increasing drag on *Skylab* as it orbited just above the atmosphere. The drag caused *Skylab* to slow in its orbit more quickly than anticipated. That slowing, along with delays in the Space Shuttle program, sealed the firey fate of *Skylab* and dashed NASA's hopes of additional missions for *Skylab* by using the Space Shuttle to boost it to a higher orbit.

With the launch of *Columbia* on the first Space Shuttle mission on April 12, 1981, though, NASA's fortunes were looking brighter. Public interest was reawakening and on Tuesday, April 14, our calculus class took a break to watch the televised landing of *Columbia* at Edwards Air Force Base in California. The student chapter of the American Institute of Aeronautics and Astronautics was revived and I joined, and met a student who was co-oping[iv] with NASA at the Marshall Space Flight Center (Marshall) in Alabama. NASA was hiring again, sometimes. I applied for a co-op position, and was told there was a hiring freeze and to try back in a year. At the end of the semester, just a few weeks later, I inquired again—my grade point average (GPA) was going up, after a couple of especially challenging courses—and was again told to try back in a year. At the end of the following semester my GPA again was higher (the possibility of working for NASA was a strong motivator!) and I sent another letter. This time there was an opening. A co-op student had transferred to the Johnson Space Center in Texas, so a replacement could be hired at Marshall. I was "on board." My dream of working for NASA was becoming reality.

• • •

IN THE 1980S, RONALD REAGAN WAS PRESIDENT and Sandra Day O'Connor was appointed the first female Supreme Court justice.

[iv] The cooperative education program is a means for students to gain on-the-job experience and earn money to help pay for their education, and is also a way for employers to evaluate prospective employees.

Smallpox was eradicated and AIDS became a health crisis. The rap music group Run-D.M.C. had a gold album, and Pac-Man became the first wildly popular video game. Space Camp was founded at the U.S. Space and Rocket Center in Huntsville, Alabama, and a movie about it debuted. The Apple Macintosh personal computer—with its mouse and other user-friendly innovations— allowed millions of people to become computer users. The hole in the ozone layer was verified to be caused by certain refrigerant compounds and use of those compounds was restricted by international agreement, the world population reached 5 billion, and leaded gasoline was banned in the U.S. to reduce environmental lead contamination. The Russian Chernobyl nuclear reactor had a meltdown, spreading radioactive material across northern Europe, the Strategic Arms Reduction Treaty (START) between the U.S. and the USSR was underway to reduce the number of nuclear weapons, and the Berlin wall fell (presaging the fall of the USSR in 1991, ending the Cold War). Iraq invaded Iran due partly to a dispute relating to oil, the Exxon *Valdez* tanker ship struck a reef in Prince William Sound in Alaska, releasing almost 42 million liters (11 million gallons) of crude oil that highly contaminated over 322 km (200 miles) of coastline, and the Space Shuttle *Challenger* exploded during launch.[20]

THE EARLY 1980S WAS A TREMENDOUSLY EXCITING TIME to be at NASA, with Space Shuttle flights becoming more frequent, *Spacelab* missions underway, and the *Hubble Space Telescope* construction nearing completion. *Voyager 1*, launched in 1977, reached Saturn, sending the first high resolution pictures of that mysterious planet. Satellites were being launched routinely, new scientific missions were in the works, and opportunities in space were broadening. On June 18, 1983 Sally Ride became the first American woman to reach space, and on August 30 Guion Bluford, Jr. became the first African-American to do so, each on board *Challenger*.[21]

May 9, 1983 was my first day at NASA, as a co-op student employee, assigned to the Design Integration Branch of the Systems Analysis and Integration Laboratory. Among other responsibilities, the Design Integration Branch supported the *Spacelab* missions by designing mockups of the experiment hardware, used for training the astronauts. Though the mockups did not actually function, control knobs and buttons, access doors, and other components the crew would use had to be accurately reproduced, including such details as the amount of force needed to open a latch or turn a knob. Every step of the procedure that the astronauts would perform needed to be accommodated by the mockup so they would be well-practiced for the mission. Using flight hardware drawings as guides, drawings were made so that sufficiently detailed mockup hardware could be fabricated. The experiment mockups were then installed in a mockup of an entire Spacelab module, located in the Payload Crew Training Complex (PCTC) at Marshall.

My first assignment was to design mockups for the *Spacelab 3* mission experiments: the Vapor Crystal Growth System, the Research Animal Holding Facility, and the Fluid Experiment System.[v] I had to learn to use state-of-the-art computer drafting equipment, but once mastered, this greatly reduced the drafting time for engineering drawings. During my first three-month co-op period I completed the drawings for these mockups and they were fabricated while I was in school the following semester. When I returned for my second co-op period, the finished mockups were ready to be installed in the PCTC module. The astronauts would train on mockups that I had designed. I was eager for more.

---

[v] The Vapor Crystal Growth System was used to grow mercuric iodide (HgI_2) crystals by vaporization and deposition. Growing these crystals in microgravity avoids strain dislocations in the crystal structure due to their weight. HgI_2 crystals are used in gamma-ray detectors (nuclear radiation detectors). The Research Animal Holding Facility provided accommodations for animals ranging in size from rodents to small primates. Physiological and behavioral responses were recorded to evaluate the effects of living in space. The Fluid Experiment System was used to grow triglycine sulfate (TGS) crystals. By growing these crystals in microgravity, convective flows are reduced, which enable more perfect crystals to be formed. TGS crystals are used in infrared detectors, and more perfect crystals could improve performance.

The *Hubble Space Telescope*, launched in 1990, is a complex, high-precision instrument that was meticulously analyzed to ensure accurate design. In 1984, I prepared detailed drawings of key portions of Hubble for use by engineers analyzing stray light paths (stray light could potentially degrade the quality of observations) and analyzing air venting during passage through the atmosphere (to ensure that pressure differentials would not exceed design levels). Though my role was quite small, it was incredibly satisfying to be part of the *Hubble* project. Despite the "*Hubble* troubles" due to a manufacturing error, it has been extremely successful and is still revealing the mysteries of deep space after 20 years of operation and, with the maintenance mission in 2009, it should provide many more years of discovery.[22]

ON JANUARY 25, 1984 PRESIDENT REAGAN ANNOUNCED the Space Station project to build a permanently-inhabited orbiting laboratory. This presented a vision for continuous, long-term habitation enabling extended experimentation and the beginnings of commercial manufacturing in space.[vi] During my third co-op period, I made drawings of several concepts that had been developed by the space station design team for the arrangement and layout of the habitat and laboratory modules. Using these drawings, the design team evaluated the ability of each configuration to meet the myriad requirements and selected the best one.

The Space Station was the project that I wanted to be part of, but I needed to complete my education. In 1985, I received my Bachelors degree, and then continued in school for a Masters degree in Mechanical Engineering. I needed only a few more courses and I completed those in December 1985. I would

---

[vi] The first commercial products made in space were 10-micron latex spheres used for calibrating microscopes. These were produced by the Monodisperse Latex Reactor on board several early Space Shuttle missions, and are sold through the National Bureau of Standards. http://nvl.nist.gov/pub/nistpubs/ sp958-lide/371-374.pdf Kornfeld, Dale M., *Monodisperse Latex Reactor (MLR) A Materials Processing Space Shuttle Mid-Deck Payload*, NASA TM-86487, January 1985, NASA, Marshall Space Flight Center.

complete my thesis while working for NASA. During the first week of January 1986, I returned to the Design Integration Branch as a degreed engineer in the Professional Internship Program. I had crossed a threshold in my career and was ready to get on with the Space Station project.

3

The Lessons of History

"If you would understand anything, observe its beginning and its development." — Aristotle

"Whoever wishes to foresee the future must consult the past; for human events ever resemble those of preceding times. This arises from the fact that they are produced by men who ever have been, and ever shall be, animated by the same passions, and thus they necessarily have the same results." — Machiavelli

"Those who cannot remember the past are condemned to repeat it." — George Santayana

ALTHOUGH PROGRESS[vii] IS NOT MADE BY MOVING BACKWARD, we can look at previous achievements for insights helpful for moving forward. One lesson is that in the course of progress there are pauses and, sometimes, periods of regression. These are natural parts of the process, when goals are clarified, new methods are developed, and support is broadened for moving forward—to cross a threshold. Such thresholds may be geographic in nature, related to technological advances, or a result of facing a new challenge.

[vii] Progress is often defined as economic growth calculated by monetary transactions and consumption. By this definition, indirect costs of producing goods or services, such as environmental degradation or human subjugation or misrepresentation, are typically ignored or are even considered to add to growth. Redefining progress to include increasing knowledge and understanding, improving quality of life, and greater long-term security and stability—as well as environmental sustainability—would provide a more comprehensive measure. Such a redefinition may also help to avoid a future that "resembles preceding times" as Machiavelli predicts, by using the knowledge of past events and current conditions, along with a broader definition of progress, to ensure that undesirable events do not repeat.

...

GREAT ACHIEVEMENTS DO NOT JUST HAPPEN, they require visions of what can be achieved, broadly felt motivation, and the means for accomplishment. Neil Armstrong's "one small step for a man" onto the surface of the Moon on July 20, 1969 during the *Apollo 11* mission was, indeed, "one giant leap for mankind," and was, perhaps, the greatest achievement in history.[viii] The vision for that achievement was presented in a speech to the U.S. Congress by President John F. Kennedy on May 25, 1961. The motivation for that achievement was the technological and political challenge posed to the U.S. by the USSR and the desire, shared by the American public who funded the effort, to meet that challenge, preferably without open, direct warfare. The means to accomplish that achievement were the hundreds of thousands of scientists, engineers, technicians, and support personnel working for NASA and its contractors, who produced the technical advances essential to accomplish such a bold vision. That vision addressed a fundamental issue of society related to a foreign challenge, which provided the motivation to see Apollo through to completion.[23]

Geographic Thresholds

THE *APOLLO 11* MISSION HAS BEEN COMPARED to another achievement, the Lewis & Clark "Corps of Discovery" expedition into the Louisiana Territory from 1803 to 1806. This is an appropriate comparison. Even in September 1806, upon Meriwether Lewis' return to Washington, D.C, one senator remarked to him that it was as if he had just returned from the Moon. Like Apollo, the expedition was a bold venture into unknown territory and it was related to addressing a foreign challenge.[24]

[viii] Other achievements are contenders for the "greatest," but the scope of cooperation for Apollo (about 400,000 people were involved in some manner), that it was an alternative to war, and that it was the first time that people set foot on a celestial body, as well as the range of "spinoffs" that were developed, together rate this a strong contender for the "greatest" ranking.

The comparison between Lewis & Clark and *Apollo 11* goes further, however. The American West was later settled by millions of people and visions of the future, by NASA and many others, include the establishment of settlements on the Moon, Mars, and elsewhere in the solar system. In the 1960s many people expected that such settlement would begin shortly after the Apollo program, as portrayed in the movie *2001: A Space Odyssey*, with orbiting hotels and cities on the Moon by the year 2001. In 1969 Wernher von Braun predicted, "By the year 2000 we will undoubtedly have a sizable operation on the Moon, we will have achieved a manned Mars landing, and it's entirely possible we will have flown with men to the outer planets." This, obviously, did not happen. Broad settlement of the American West, though, did not begin immediately upon the return of Lewis & Clark, either. The presence of the native peoples was one factor, but, more consequentially, the vision for settlement was not sufficiently developed, and the motivation and the means did not yet exist to support such a vision.[25]

At the time of the Corps of Discovery, most Americans lived within 80 km (50 mi) of the Atlantic coast, and the region between the Ohio River and the Great Lakes was considered the "northwest territory" being opened to settlement. The vision for development of the region west of the Mississippi River took many years to become widespread, and the "manifest destiny" of the U.S. to span the continent did not acquire that name until the 1840s. But during the early 1800s, certain key technologies were developed that would provide the means for later major settlement. These technologies related to transportation, communication, and agriculture.[26]

Railroads could provide the means for easy, rapid, and economical commerce across the continent, which was Thomas Jefferson's dream even before the Louisiana Purchase in 1803. On July 1, 1862, while the U.S. Civil War still raged, President Abraham Lincoln signed the Pacific Railway Act, authorizing the first transcontinental railroad to cross the heart of the West. The Act provided guarantees—of money and land—to the Union Pacific and Central Pacific railway companies, reducing their financial risk to build the railroad. This was an attractive incentive and on May 10, 1869, less than 7 years from the signing and after 2,859 km (1,777 miles) of track were laid, a golden spike ceremonially joined the tracks from Sacramento, California and Omaha, Nebraska at Promontory Point, Utah. This allowed transcontinental rail travel that greatly reduced the time required to cross the country, and on June 4, 1876 the *Transcontinental Express* passenger train arrived in San Francisco only 83 hours and 39 minutes after leaving New York City—a journey that previously was measured in months. The speed of long-distance communication was increased even more dramatically by the telegraph lines that were extended along with the railroads, sending signals at close to the speed of light.[27]

Advances in agricultural technology were also critical. Zebulon Pike, in his 1810 report on his exploration of the southwest, described much of the Great Plains as a desert—a dry, treeless region unsuitable for agriculture—and with the technology of the early 1800s the prairies could not be farmed—the thick sod was essentially impenetrable by the wooden and cast iron plows of the time. Thus, the invention of steel plows in the 1830s was an especially significant advance. The early 1800s also saw such incremental and breakthrough advances as reapers, threshers, harrows, and canning of food for long-term preservation. Later important developments include barbed wire fencing, windmills to power pumps to draw groundwater from wells, and irrigation techniques to allow farming of dry regions. These technologies enabled settlers to provide for themselves, and to produce excess goods that could be sold and delivered eastward by the railroads.[28]

HAVING CLEAR BOUNDARIES ALSO PROMOTED SETTLEMENT by reducing risks, and the geographical boundaries were set by the Oregon Treaty of 1846 that established the boundary between the U.S. and British Canada at the 49th parallel west of the Great Lakes, and the Mexican-American War (1846 to 1848) that added the southwest region to the U.S. After the boundaries were established, tens of thousands of people headed west. Yet even then, the goal was not to settle in the Great Plains, but to reach

California, Nevada, or Oregon, where gold, silver, or fertile fields suitable for standard farming techniques could be found, or, for the Mormons, to reach Utah where respite from religious persecution awaited. The Great Plains were still largely left to the native peoples.[29]

In addition to the Pacific Railway Act, other legislation was instrumental in settlement of the West. The Homestead Act, also enacted by Lincoln in 1862, provided inexpensive land for individuals or families to be farmers, and allowed new immigrants, farmers without their own land, women, and former slaves an opportunity to acquire 65 hectare (160 acre) plots of land to call their own. Landholders were only required to be the head of their household, to be at least 21 years old, and to pay fees totaling $18. In order to receive the "patent" of ownership, the homesteaders had to build a home and farm the land for five years. By the time the Homestead Act was repealed in 1976, about 109 million hectares (270 million acres) (10 percent of the area of the U.S.) were claimed and settled under this Act. The Mining Act, enacted in 1872, further encouraged development by declaring that public lands where minerals could be mined were free and open to exploration. The Mining Act effectively subsidized extraction of the resources of the West, allowing prospectors and companies to mine for hard-rock minerals for only $5/acre. These Acts and technological advances reduced the risks of settling the West.[30]

BY THE 1860S, THE VISION FOR THE U.S. TO EXTEND across the continent was widely held, and the means for development were in place with the construction of the railroad and telegraph (enabling infrastructure), invention of the steel plow and other agricultural equipment (critical technologies), and passage of the Oregon Treaty, Homestead Act, and Mining Act. (supportive legislation). Even so, major migration would not have occurred without strong incentives for the settlers to uproot their lives and endure the travails of settling a land considered by many people to be untamed and dangerous.

Immigration from Europe was ongoing throughout early U.S. history, but the wave of millions of immigrants began rising in the

1840s. Many were relegated to living in crowded urban tenements or other segregated communities in eastern cities, and the influx of diverse cultures was a fundamental issue of the mid-1800s. Integrating the immigrants into American society was not easy. Many were from non-English-speaking countries and they were frequently discriminated against, and encouraged to go west. The growing industrialization of the U.S. was another fundamental issue. With an ever-increasing need for raw materials and energy, the mines in the East could not keep up with demand. The Mining Act opened the West to exploration and extraction of resources.[31]

The motivation for moving west acquired broad support with the growing immigrant population, the need for raw materials, and the desire for a fresh start for survivors of the bitter rivalry of the U.S. Civil War. That motivation—converging with the new technologies, the easier access, and the decline of the native peoples due to disease and conflict—resulted in millions of people settling along the rivers and rolling hills of the prairies, in the foothills of the Rocky Mountains, and throughout the West. Following the Civil War, the migration occurred so rapidly that with the 1890 census the U.S. Census Bureau declared the West settled. The rapid rate of change during this period was reflected in the agricultural output of the U.S., which more than doubled from 1870 to 1900, largely because of the new farms in the West and farming practices that produced higher yields. The West was no longer the distant, harsh desert it had been considered only a few decades earlier. To millions of Americans, it was home.[32]

Key Points:

1. Three general factors are important for an achievement: vision, motivation, and means.

2. For major societal achievements the vision must address important issues, the motivation must be broad-based, and the means—consisting of enabling infrastructure, critical technologies, and supportive legislation—must be in place.

3. These factors may take decades to develop following a seminal event, and prior to the maturing of them the achievement may seem to be far off, but when they mature and converge the achievement will occur rapidly.

4. Clear boundaries (geographic and legal) help to reduce the risk for performing particular actions.

5. Government incentives can reduce risks, encouraging individuals and companies to perform desired actions.

6. Legislation is key to promoting a vision, clarifying boundaries, and providing incentives for individuals and companies.

7. A range of technologies were important for settlement of the West. Some were breakthrough and some were incremental, but many were developed after settlement began, when the specific needs became clear.

Technological Thresholds

THE DEVELOPMENT OF PRACTICAL AIR TRANSPORTATION crossed a threshold that provided a new perspective on the world and, by providing rapid transportation between distant locations, had the effect of bringing nations closer together. For the most part the advances came incrementally, over many years and from many different people. The Wright brothers—the famous bicycle shop owners who flew the first successful airplane—built on the efforts of many, from Leonardo da Vinci who designed helicopters and ornithopters (with flapping wings) in the 15th and 16th centuries to Sir George Cayley who determined in 1799 that the wings could be fixed in place to Otto Lilienthal who built and tested gliders in the late 1800s to Octave Chanute who sponsored aeronautics experiments in the 1890s. The Wright brothers' interest in flight was sparked by a flying toy given to them by their father, and grew by reading his books on flight in nature. In 1896, while Orville was recovering from typhoid, they read that Lilienthal had died in an accident while testing a glider of his own design. By 1899, their interest led Wilbur to contact the Smithsonian Institution to request any aeronautical research information available. They were young men, in their 20s, who wanted to fly and to make their mark on the world. Toward that end they embarked on their own research into airfoil design, control methods, and propulsion. Their success made a dramatic mark, indeed, but it was not certain.[33]

In the late 1800s, a number of aviation enthusiasts were making incremental advances on all aspects of aeronautics and converging on a successful design. In 1894, Chanute published *Progress in Flying Machines*, bringing together in one volume much of the state-of-the-art knowledge of aeronautics, and James Means published *The Problem of Manflight*, as well as his *Aeronautical Annual*, which included an 1896 article comparing the control and balance issues of flying with bicycling. These publications coherently described the current knowledge and clarified the remaining challenges.[34]

Also in the 1890s, Samuel P. Langley, secretary of the Smithsonian Institution, first built a steam-powered model airplane that traveled about 1.2 km (¾ of a mile), and then a larger gasoline-powered model. In 1898, he received a grant of $50,000 from the U.S. War Department to build a piloted airplane. In 1903, he was ready to test his airplane with a pilot on board, and on December 8, the airplane was catapulted from a houseboat on the Potomac River. While climbing, the aft wing broke off and the vehicle crashed, but even if the structure had held, the airplane would not truly have flown. Though the engine was powerful enough to propel the craft, inadequate controls left it at the mercy of the winds.[35]

True flight involves more than simply moving through the air or staying aloft. In addition to sufficient lift and thrust, control of attitude and orientation is essential. Perhaps influenced by Means' article, as well as their own experience with bicycles, what the Wright brothers realized, that had not been fully appreciated by other aeronautics researchers, is that to turn in flight airplanes need to be controlled along the roll axis, as well as the pitch and yaw axes.[ix]   They provided that control by mimicking the control method of birds, designing their vehicles so that the wings, though fixed in place, could deform in a precise manner—a technique called wing warping. This seemingly simple insight was key to providing controlled banking, to make graceful turns in the air or to maintain a heading even with a cross-wind.

After constructing their first unpowered glider the Wright brothers contacted the Weather Bureau for information on locations with strong, steady winds and chose Kitty Hawk, North Carolina, as their test site. When their gliders did not perform as expected, they found that the lift coefficients in their equations were inaccurate. The coefficients were from a table of lift vs. angle of attack[x] that Lilienthal had developed, and the errors may have contributed to his death. To determine the correct values, the Wrights built a wind tunnel and made about 200 models of wing shapes to test. With correct data they were then able to accurately calculate the performance, and after verifying their calculations with the glider they designed and built their first powered airplane, the *Wright Flyer*.[36]

The first successful flight, of 37 m (120 feet), on December 17, 1903, was a tremendous achievement, but news spread slowly, perhaps due to skepticism related to highly publicized attempts by Langley and other aviation enthusiasts. The first published report was in the March 1, 1904 issue of *Gleanings in Bee Culture*, hardly a major news publication. Though the fundamental issues of

---

[ix] If you draw a line from the nose of an airplane to the tail, *roll* is the rotation about that line, i.e., banking with left wing high-right wing low and vice versa, controlled by wing-warping by the Wrights' and by ailerons on most airplanes today. If you draw a line from wingtip to wingtip, *pitch* is the rotation about that line, i.e., nose up or nose down, controlled by the elevator (usually in the tail assembly, though for the Wright Flyer the elevator was in front). If you draw a line from the roof of an airplane through the belly, *yaw* is rotation about that line, i.e., nose left or nose right, controlled by the rudder.

[x] The *angle of attack* is the angle between the chord line of a wing (drawn through the trailing and leading edges across the wing width) and the relative direction of the wind. In general, the greater the angle of attack, the greater the lift, but also the greater the drag.

controlled flight were solved, further advances were slow in coming. It was not until September 20, 1904 that the Wrights flew in a complete circle for the first time, returning to the starting point. In 1908, almost 5 years after the first flight, Glenn H. Curtiss won a trophy from Scientific American for flying an airplane a distance of 1 kilometer (2/3 mile).[35, 37]

Progress was slow indeed in the first few years, but during the first decade of flight many critical technological advances were made such as enclosed cockpits that greatly increased comfort for the pilot and passengers, monocoque[xi] fuselage construction that lightened the structure while simplifying assembly, and the use of wingflaps and ailerons[xii] to improve control in flight. These advances were also all developed in European laboratories. Due to these advances, and because of patent disputes between American aviators and the Wrights, European aviators were overtaking early U.S. leadership.[38]

In response to the competition posed by the Europeans, and the prospect of being involved in World War I, the National Advisory Committee for Aeronautics (NACA) was created by Congress and signed into existence by President Woodrow Wilson on March 3, 1915. Its mission was to "direct and conduct research and experimentation in aeronautics, with a view to their practical solution," and it was involved in all areas of aeronautics. The facilities constructed by the NACA, such as wind tunnels, were used to perform research and development beyond the capabilities of private companies. In addition, the NACA recommended creation of the Manufacturers Aircraft Association in 1917 to

---

[xi] *Monocoque* is a French word meaning "single shell." It is a construction technique in which the external skin, rather than an internal framework, supports the bending and twisting stresses.

[xii] *Wingflaps* are moveable control surfaces along the trailing edge of a wing that increase lift and reduce the stalling speed, allowing for landing at slower speeds than otherwise possible. They operate together on each wing.

Aileron is a French word meaning "little wing." Ailerons are the moveable control surfaces along the trailing edge of a wing that provide roll control. They operate in opposite directions on each wing, with the aileron on one wing moving up while the aileron on the other moves down.

promote cross-licensing of aeronautics patents to resolve bitter patent disputes and recommended inauguration of regular airmail service in 1918 to promote commercial aviation.[39]

FOLLOWING WORLD WAR I, the Curtis JN-4 airplanes (Jennys) used by the military were no longer needed and many were sold to the pilots who had flown them during the war, who then traveled the country selling rides and putting on "barnstorming" exhibitions of aerial derring-do. Especially during the 1920s, barnstorming— the first major form of civil aviation—became one of the most popular forms of entertainment and brought the experience of flying to millions of people, as observers if not as passengers.[40]

By the late 1920s, the success of the NACA at regaining U.S. leadership was demonstrated by the defining feat of that bold and brash decade: Charles Lindbergh's solo non-stop crossing of the Atlantic Ocean from New York to Paris, in May 1927, in his airplane, *The Spirit of St. Louis*, that incorporated advances from the NACA in its design. For his efforts, Lindbergh won the $25,000 Orteig prize, sponsored by New York hotel owner Raymond Orteig to promote the advance of aviation. The flight captivated the public and Lindbergh was greeted by huge crowds in Paris and treated to a ticker-tape parade upon his return to New York City. The feat was also immortalized with a new dance, the Lindy Hop, an exuberant expression of freedom. Lindbergh's "hop" was a long way from the Wright brothers' first flight 24 years earlier, but it showed the potential for future developments.[41]

During World War II, the NACA focused more on solving specific problems than on advancing general aeronautical knowledge. One major advance, however, was the development of the laminar-flow airfoil, that eliminates turbulence at the wing trailing edge that had limited aircraft performance. It was apparent, though, that propeller-driven aircraft were approaching the limits of their capabilities and that another method of propulsion would be needed. The answer lay in Newton's Third Law of Motion, "for every action there is an equal and opposite reaction." Rockets rely on this principle, but another approach, using atmospheric oxygen during combustion, was conceived in 1928—the jet engine. The concept was developed independently in England and in Germany (in the 1930s), but it was not used to power an airplane until World War II, first by Germany, then by England. Technical issues for these jets limited engine life to about 10 hours due to metal fatigue, related to the use of steel rather than chromium and other rare metal alloys which were scarce during World War II. Following the war better materials were used that extended engine life, broadening the practical applications and leading to the first commercial jet airliners: the DeHaviland *Comet* in 1952 and the Boeing *707* in 1958.[42]

THE FORMATION OF THE NACA BY CONGRESS was an enormously successful response to the technological advances of the European aeronautics researchers. As described by Arnold Levine, "Three features of NACA practice may serve to make its success comprehensible. Its mission made it complementary to, not competitive with, the [military] services and industry; its research was only loosely coupled to its users; and its laboratories enjoyed a certain autonomy in the selection of specific research projects and the manner in which research would be conducted." These features provided a work environment receptive to new ideas with the ability to pursue them with scientific rigor, stimulating the creativity of the NACA engineers and scientists. The results of the NACA research projects, documented in thousands of technical reports, led to dramatically improved performance and safety of airplanes, and laid the groundwork for future efforts in space.[43]

Airplane manufacturers, government agencies, and private supporters all played important roles in the development of reliable, rapid, affordable, and safe air travel. By 1958, 55 years after the Wright brothers' first flight and 31 years after Lindbergh crossed the Atlantic Ocean, the infrastructure of airports and the air traffic control system were in place, the legislation and regulations defining the roles and responsibilities regarding aviation were implemented, and the technology of jet airplanes had superseded piston-engine aircraft. Aviation was poised to fill the desire for improved travel and the need for greater interaction between nations. Especially with the development of commercial jet

airplanes—that greatly reduced the time and cost to travel—
aviation had its "transcontinental railroad." Aviation "took off"
and increasing millions of people each year took the opportunity to
visit distant places and foreign countries.[44]

Key Points:

1. A clear understanding of the challenges is important in order
   to overcome them, and technical excellence is key to doing
   so.

2. Incremental advances help "set the stage" for breakthrough
   developments.

3. Researchers are most productive when involved with
   choosing the type and manner of research undertaken.

4. Government support can help private enterprise to succeed by
   providing information or services not otherwise available and
   by serving as a customer while a market develops.

5. Congressional action can effectively respond to important
   issues.

6. Recreational activities can be important in developing a
   commercial market.

7. Awards and monetary prizes effectively promote specific
   advances.

Thresholds of Challenge

FRUIT TREES HAVE PLAYED SURPRISING ROLES in the development
of space flight. In 1666, when Cambridge, England had succumbed
to the plague sweeping across Europe, Isaac Newton went home to
his grandmother's farm in Lincolnshire. While there, he watched
an apple fall from a tree. This common event, along with the
uncommon ideas of Nicolaus Copernicus, Galileo Galilei,
Johannes Kepler, and others, led him to conceive of universal

gravitation[xiii] as a force that determines the movement of objects through space, including planets, comets, and spacecraft; and to articulate his laws of motion[xiv] that define the relationship between force and motion, critical knowledge for designing rockets.[45]

Robert Goddard in the U.S., Konstantin Tsiolkovsky in Russia, and Hermann Oberth in Germany are considered the "fathers" of modern rockets. Working independently, and unknown to each other, they developed the basic design for liquid-fueled rockets, using the theories of Newton. However, they were each influenced by the visions of writers such as Jules Verne (especially his story *From the Earth to the Moon*) that stirred their imaginations. Goddard had his own vision, as well, on October 19, 1899, while pruning dead branches from a cherry tree in Worcester, Massachusetts. As he later described the occasion: "It was one of the quiet, colorful afternoons of sheer beauty which we have in October in New England, and as I looked toward the fields at the east, I imagined how wonderful it would be to make some device which had even the possibility of ascending to Mars, and how it would look on a small scale, if sent up from the meadow at my feet." He was 17 at the time, and that vision set the course for the rest of his life. It took many years for Goddard to demonstrate the validity of the basic concepts, but he succeeded on March 16, 1926 when he test-fired a gasoline and liquid oxygen-fueled rocket that reached an altitude of 13 m (41 feet). His later rockets launched

---

[xiii] Every mass attracts every other mass by a force proportional to the product of their masses and inversely proportional to the square of the distance between them. In equation form this is: $F = G (m_1*m_2)/r^2$, where F = attractive force, G = gravitational constant (the proportionality factor), m_1 and m_2 = the masses being considered, and r = the distance between the masses.

[xiv] Newton's Laws of Motion:

1. Inertia - An object at rest tends to remain at rest and an object in motion tends to move in a straight line at a constant speed, unless acted upon by an external force.

2. Acceleration and force - The change in motion (a, acceleration) of an object (of mass, m) is proportional to the force applied (F). In equation form this is: $F = m*a$

3. Reaction - For every action, there is an equal and opposite reaction.

barometers and cameras much higher into the sky, at speeds up to 885 km/h (550 miles/h).[46]

In 1919, Goddard wrote "A Method of Reaching Extreme Altitudes" about using rockets for upper-atmosphere research. In that paper, he developed the mathematical theories of rocket propulsion and he also proposed sending a spacecraft to the Moon that would explode flashpowder to announce its arrival. His ideas were not well received, though, and he was criticized in the popular press and by other scientists, and given the moniker "loony Goddard." He was too far ahead of his time for them to understand, but rocket engines are among the most efficient means of converting combustion to work—raising the mass of a rocket into the sky—and their day was coming.[47]

Even before Goddard, in 1903, Tsiolkovsky had published *The Exploration of Outer Space by Means of Reaction Apparatus*, describing the state of weightlessness, the theoretical function of rockets in a vacuum, and rockets using liquid fuel (liquid oxygen and hydrogen). He also presented his equation for rockets,[xv] based on Newton's Third Law of Motion. Earlier, in 1895, while visiting Paris and seeing the Eiffel Tower, Tsiolkovsky hypothesized extending its height into space, to a "celestial castle" at geosynchronous orbit. At that altitude (35,800 km, 22,240 mi), its orbit would be synchronized with the Earth's rotation and it would appear, from the Earth, to be stationary.[xvi] Materials of the time were woefully inadequate to construct such a tower, though, so he continued developing his ideas on rockets. Also in 1895, he published *Dreams of Earth and Sky*, a novel that includes a vision of settling space, using asteroids and other bodies in space as sources of materials. In that novel he described space suits, space stations, and the use of solar energy. In recognition of his

---

[xv] In its simplest form "Tsiolkovsky's equation" is: $\Delta v = v_e \ln(m_o/m_1)$ This expresses the relationship between the initial total mass (m_0), the final total mass (m_1), and the velocity of the exhaust (v_e), with the change in vehicle velocity (Δv). "ln" means the natural logarithm.

[xvi] Communications satellites are positioned at geosynchronous altitude so that Earth-bound receiver dishes can be aimed and locked into position.

importance to the development of space travel, the USSR launched *Sputnik I* near the centennial of Tsiolkovsky's birth. [48]

In the early 1900s, Oberth was also studying rocketry and, in 1922, he submitted a doctoral dissertation that was rejected as being too speculative. He then published it as *By Rocket into Planetary Space*. His theoretical analyses were the most comprehensive at the time, stimulated German experiments in liquid-fueled rocketry, and led to the formation of the German Society for Space Travel. In 1929, German film-director Fritz Lang made a silent movie, *The Woman in the Moon*, in which a trip to the Moon is portrayed with considerable technical accuracy, including showing a multi-stage, liquid-fueled rocket that had to achieve the correct velocity to reach the Moon. In a strange foretaste of the *Apollo 13* mission in 1970, there was even a rupture of an oxygen tank during the journey, but perhaps the most surprising prescient aspect is the first use of the countdown ("3, 2, 1, liftoff"), unlike any previous description of a rocket launch. The high degree of technical accuracy was due to Oberth being an advisor for the film. As a publicity stunt, Oberth was to build an actual rocket to be launched at the premiere showing. He was aided by a team including 18-year-old Wernher von Braun. They constructed and tested a small rocket engine on July 23, 1930, but were unable to complete the rocket in time for the release of the movie. This collaboration between Oberth and von Braun, however, was the start of a lifelong friendship.[49]

The exigencies of war result in receptiveness to new ideas when an advantage can be obtained and rocket development was one such idea. During the Great Depression of the 1930s, the German Society for Space Travel disbanded, but with the onset of World War II, Nazi Germany enlisted some of its members in the military effort. The infamous Peenemunde rocket development center was built under the technical direction of von Braun, creating rockets that could fly hundreds of kilometers (miles), landing as bombs on England. These were the V-2 rockets, the first long-range ballistic missiles, incorporating advances in propulsion, manufacturing, and GN&C[xvii], that would be built upon for later rocket vehicles. After World War II, von Braun and many of the engineers and scientists who worked with him were brought to the U.S. to develop improved rockets for the Army. During this time the USSR was also actively pursuing missile development, utilizing the German rocket scientists that they had captured during the war. The seeds of the space race were planted.[50]

The U.S. planned to launch satellites as part of the International Geophysical Year (IGY) in 1957 and 1958, expecting to be the first. The launch of *Sputnik I* by the USSR on October 4, 1957, using a ballistic missile, "scooped" the U.S. *Sputnik I* was rather small by today's standards, weighing only 83 kg (184 pounds), but as it circled Earth emitting electronic beeps it showed the potential for ballistic missiles launched from protected sites deep inside the USSR to deliver bombs on targets anywhere in the U.S.[xviii] One month later, on November 3, 1957, *Sputnik II* was launched, weighing 508 kg (1,120 lb) and carrying Laika, a dog. This turn of events provided tremendous motivation for the U.S. to respond rapidly. The first few attempts, with the Navy-sponsored *Project Vanguard*, proved disastrous, exploding during launch. Von Braun's Army team then proceeded with their plan to launch a satellite, and on January 31, 1958, *Explorer 1*, weighing 14 kg (31 lb), was successfully sent into orbit by the *Juno 1* rocket. In response to the challenge by the USSR, the National Aeronautics and Space Administration (NASA) was formed by an act of Congress on October 1, 1958, absorbing the NACA, as well as the Jet Propulsion Laboratory and groups from the Naval Research Laboratory and the Army Ballistic Missile Agency. New field centers were formed, including the Marshall Space Flight Center in Alabama, where von Braun was named director. The space race between the USSR and the U.S. was "launched."[51]

[xvii] GN&C refers to guidance, navigation, and control.

[xviii] This concern was reinforced by the announcement from the USSR a few days later that they had exploded a hydrogen bomb. "New H-Bomb Is Exploded By Russians," Associated Press, October 7, 1957.

UNDER THE TECHNICAL LEADERSHIP OF SERGEI KOROLEV, and utilizing the theoretical foundation provided by Tsiolkovsky, the USSR was able to reach space ahead of the U.S., but *Explorer 1* did more than simply emit electronic beeps as had *Sputnik I*, it made a discovery that highlighted the importance of the scientific study of space and revealed some of the hazards found there: the deadly radiation of the Van Allen belts around the Earth.[52]

Following the launch of Russian Yuri Gargarin into orbit on April 12, 1961 by the USSR, a dramatic response was sought by the U.S. and, so, NASA was assigned the defining feat of the 1960s—landing men on the Moon. On May 25th, President Kennedy presented the vision for the Apollo project near the end of a speech to Congress on national security, saying, "I believe that this nation should commit itself to achieving the goal, before this decade is out, of landing a man on the moon and returning him safely to the earth. No single space project in this period will be more impressive to mankind, or more important for the long-range exploration of space; and none will be so difficult or expensive to accomplish. ... in a very real sense, it will not be one man going to the moon—if we make this judgment affirmatively, it will be an entire nation. For all of us must work to put him there."[53]

The Apollo program was undertaken to unequivocally demonstrate the technical superiority of the U.S. and, as Kennedy predicted, the achievement was "impressive to mankind." The success was astounding and, at the time, it seemed that the pace would continue as NASA developed plans for more missions to the Moon and for permanently inhabited space stations, as well as missions to Mars.[54]

The frantic pace of the 1960s space program included many firsts by the U.S., as well as the USSR (summarized in endnote 55). In the 1970s, the pace of space efforts changed as fewer, though more sophisticated, exploration spacecraft were launched and the first space stations were occupied. The commercial space industry was developing and an increasing percentage of launches was for communications and Earth observation satellites that could be supported by sales of services. The *Apollo/Soyuz Test Project* in 1975 was notable for the first international handshake in space when Thomas Stafford and Alexei Leonov, commanders of the respective spacecraft, greeted each other in orbit around the Earth. That handshake reassured the world that despite the Cold War between these two superpowers activities in space would be peaceful.[55]

Considerable technical advances were made in the fields of materials science, computers and controls, health and medicine, and much more. Extensive facilities built to enable these missions include not just the launch and tracking stations, the manufacturing facilities, research and development laboratories, and operations centers on Earth, but also satellites to aid communication and tracking and to observe the weather on Earth and in space. The technical advances and the infrastructure provide benefits far beyond their initial purposes.[56]

Among the greatest benefits of Apollo, though, was the broadly felt attitude that with resoluteness and creativity, the most intractable problems could be solved. "We can send a man to the Moon, why can't we ...?" became part of our lexicon, expressing the desire for other problems in society to be addressed with as much determination as the Apollo program. In the long tradition of American ingenuity and innovation, NASA was the epitome, and the effects of NASA's spirit were felt throughout society.

Key Points:

1. Inspiration is vital to develop a vision, as shown by the early rocketry pioneers who were inspired by Jules Verne and other writers.

2. Receptiveness to new ideas is key to obtaining advances.

3. The achievements of the early space age were the result of joint government/private industry efforts, with the government taking the primary role of defining the missions, organizing and coordinating the efforts, and providing the funding.

4. Until a market is developed, the payoff to private enterprise may have to come from government contracts rather than "market share."

5. International coordination of exploration and research (such as the IGY) can provide a greater knowledge return than may be obtained otherwise.

Thresholds and Change

OVER THE COURSE OF A CENTURY, or less, great changes can occur, representing the crossing of numerous thresholds. Consider the changes that occurred over one lifetime.

On August 1, 1883, President Chester A. Arthur pulled a silken cord that set in motion the machinery powering the largest display of electric lights at the time. The occasion was the opening of the Southern Exposition in Louisville, Kentucky. According to the Filson Historical Society the exposition was a way to showcase that "Louisville was ready to claim her place in the trade market" with exhibits, lectures, concerts, and weekly fireworks displays. This was one of many such expositions held in numerous cities in the late 1800s, but what made this exposition unique was the extensive use of electric lighting, perfected by Thomas Edison a few years earlier. Edison himself went to Louisville to install the generators and the power distribution system to illuminate the buildings and grounds. With 4,600 electric lamps lighting the main building, this was the first successful nighttime exposition. Over 770,000 people attended during the first 88 days, and the planned 100-day exposition was extended into 1887. "Electric lighting allowed for late afternoon and evening entertainment with the evening highlight being the illumination of the lights as the sun set. The Southern Exposition marked the beginning of a new industrial era for Louisville and other cities of the South."[57]

The year before that exposition, my grandfather, David Wolf, was born in Louisville. It was a time of coal-powered industry and steam locomotives, horse and buggy transportation, gas lighting and kerosene lamps, coal or wood for heat, and door-to-door ice deliveries for residential iceboxes. Black and white photography was well-refined, and there are photographs of my grandfather as a child with his family. The telephone was a new invention, just beginning to become an indispensable device. However, there were no automobiles, no radios, no televisions, no movies, no airplanes, no computers, no air conditioners, and certainly no spacecraft. The American West was in the prime of its "wild" period and Buffalo Bill Cody was forming his Wild West show. In most states women were not allowed to vote or hold office, diseases such as smallpox and polio caused epidemics regularly, and most Americans lived on farms. As a young man, my grandfather worked in a shoe factory. After marrying, he opened his own shop and repaired shoes while my grandmother sold "notions" (small items). During the Great Depression and World War II, when new shoes were not so easily obtained, they earned a good living, but never became wealthy.

By the time of my grandfather's death in September of 1969, however, the West had long been "tamed," women had full legal rights in the U.S., smallpox was close to being eradicated and polio was greatly reduced, the vast majority of Americans lived in cities or their suburbs, horse-riding was largely a recreational pursuit, electric lights and automobiles were ubiquitous, industry was mostly powered by electricity, railroads were driven by diesel-electric locomotives, long-distance passenger travel by train had largely been supplanted by airlines, and the U.S. had landed men on the Moon. The amount of change that occurred over my grandfather's 86 years was phenomenal.[58]

The 21st century will likely experience the same degree of change. Such change always occurs in response to the key issues of the era. The direction and rate of change often depend upon public policy, as well as technological developments. However, the consequences of policy actions taken in response to key issues are not always the ones intended. The unintended consequences may be good or bad, small or great, immediate or delayed, but they always occur. To minimize undesirable consequences of policy actions, it is important that the policies clearly address key issues, and that they be carried out in a manner consistent with the vision they support (i.e., the end does not justify the use of any means to achieve it that is not consistent with the desired end result). Even with the best preparation, however, there are always risks.

4

Risk: The Surly Bonds of Earth

"To dare is to lose one's footing momentarily. To not dare is to lose oneself." — Soren Kierkegaard

"I'm not afraid of storms, for I'm learning how to sail my ship." — Louisa May Alcott

THRESHOLDS ARE THE TRANSITIONS between the familiar and the unknown. Crossing thresholds, therefore, involves a certain degree of risk. But, risk is inherent in any activity, even getting out of bed in the morning, or staying in bed. We cannot avoid risks, though with careful preparation risks can be minimized. Even with the best efforts, mistakes or poor judgment or unanticipated events can occur that result in setbacks or failure. When entering new territory we often learn as we go, making mistakes in the process, and that is as true of entering space as it has been of entering any new territory on Earth. How we deal with mistakes and failures is often more important than the events themselves.

• • •

JANUARY 1986 IS A MONTH THAT I WILL NEVER FORGET, nor will an entire generation. The first week of January I returned to work in the Design Integration Branch and I was settling into my new position as an intern. With all of the activities in progress it was a very exciting and busy time, and Shuttle launches were becoming more frequent, almost routine. I had watched several launches on television and saw one in person (*Challenger* on mission STS-8, the first nighttime Shuttle launch on August 30, 1983). On January 28, my office-mates and I were at our desks working rather than watching the launch of *Challenger* on mission 51-L. After the launch a coworker came into the office with a kind of drained expression and said there had been an accident and *Challenger* had been destroyed. He was known for his dry sense of humor and at first we thought—hoped, even—that it was some kind of twisted joke, but then realized that it wasn't. *Challenger* and her entire crew were lost during launch. The image of her destruction was burned into the memories of the schoolchildren who watched the launch on TV as it happened. The first "citizen-astronaut," Christa McAuliffe, a teacher, was on board and the mission was intended to educate and inspire students about the space program and science. Instead, it was a lesson on the risks of space travel and on the effects of poor decisions. We knew the consequences would be severe, but none of us would have predicted that the next Shuttle mission would not fly for two-and-a-half years.

In his address to the nation on the day of the *Challenger* accident, President Reagan expressed the shock and pain of the loss and talked of the risks inherent in exploration and discovery. Invoking the words of John Gillespie Magee, an airman killed in World War II, he said, "The crew of the space shuttle *Challenger* honored us by the manner in which they lived their lives. We will never forget them, nor the last time we saw them, this morning, as they prepared for their journey and waved good-bye and 'slipped the surly bonds of earth' to 'touch the face of God.'"[59]

A TRAGEDY SUCH AS THE CHALLENGER ACCIDENT CAN, after the initial shock wears off, stimulate greater efforts to press ahead, while ensuring that a repeat never happens. An intense investigation began shortly after the accident to identify the cause and determine how to prevent its recurrence, and I had an opportunity to be part of that effort. Interns were required to work in two groups other than their home group for three-month rotations to obtain a broader perspective on the roles and responsibilities of the agency. One option was to work in the Solid Rocket Booster (SRB) Project Office, which I did in the spring of 1986. The investigation into the cause of the *Challenger* accident was in full force. The design was thoroughly reassessed and every possible cause considered. To determine whether errors had been made during manufacturing, every step of the process was evaluated. I was assigned to review the labor time required to make each SRB (the "touch labor" hours) to look for any trends or indications that might provide insight into the SRB failure. One way to do this was to track the "learning curve" or reduction in hours required to fabricate successive SRBs. The touch labor hours for manufacturing each SRB had been meticulously recorded, so it was a matter of converting that data into the appropriate form for analysis, which might provide insight into the failure.[60]

Learning curves apply to mass production and are based on the proclivity that as more copies of an item are manufactured less time is needed to make each one until an optimum time is reached. It is generally assumed that the reduction is basically a straight line. When producing thousands or tens of thousands of an item this is often the case, especially with short cycle time items—those that can be manufactured in a matter of minutes or hours. For large items that are produced in smaller quantities (a few hundred or so) and require weeks or months to complete (long cycle) the situation is somewhat different. Due to increased complexity there is a slower rate of learning for the first few items, followed by a more rapid reduction in labor hours for later items until approaching the optimum (standard) time. When plotting time vs. copy number the result is an S-curve, which may be a shallow curve (almost a straight line) or a very pronounced curve. In either case (straight line or S-curve), as more copies are produced the cost per copy tends to decline.[61]

Design changes and manufacturing problems can greatly affect the number of labor hours required to make each copy and this was certainly the case for the SRBs. The number of touch labor hours for each SRB was following a distinct S-curve until there was a design change or an accident, such as a fire in the propellant mixing pit. Following those events the number of hours jumped to near the level of the first SRB. As production continued the decline was more rapid and, after several more SRBs were made, the curve would approach the point where it might have been without the disruptive event. The result, though, was a significant increase in the time required to manufacture a number of SRBs and something critical may have been forgotten. As it turned out, although mishaps had occurred during manufacturing none were related to the *Challenger* accident. At the end of my three-month rotation to the SRB Project Office I turned in my report, concluding my role in the *Challenger* investigation. Later the *Challenger* Commission determined that the immediate cause of the accident was insufficient flexibility of the O-ring seals joining the SRB segments—due to cold temperatures at the time of launch—that allowed hot gases to burn through the side of an SRB.[xix][62]

INSTEAD OF BEING INSPIRING LESSONS in science, the activities planned around that mission and the demonstrations that McAuliffe was to perform in space on "The Ultimate Field Trip" became reminders of the tragedy. But out of tragedy can come the determination to carry on. The widow of the commander of mission 51-L, June Scobee, along with family members of the other astronauts on board *Challenger*, "resolved to create a living memorial to the *Challenger* crew" and established the "first interactive space science education center," founding the *Challenger* Center for Space Science Education on April 24, 1986. By 2009, there were over 50 *Challenger* Centers in the U.S., Canada, and England, with more planned. Thousands of students have participated in *Challenger* Center programs and activities, embodying the spirit of the 51-L mission, and preparing for the future.[63]

[xix] Report to the President By the Presidential Commission on the Space Shuttle Challenger Accident," June 6, 1986, Washington, D. C., also known as "the Rogers Report" since the Commission Chairman was William P. Rogers, http://history.nasa.gov/rogersrep/genindex.htm

48

Part II

Envisioning the Future

"It is difficult to say what is impossible, for the dream of yesterday is the hope of today and the reality of tomorrow." — Robert Goddard, 1904[xx]

• • •

IT CAN ALSO BE DIFFICULT TO SAY WHAT IS POSSIBLE. Predictions from the 1950s of electricity from nuclear power plants that is "too cheap to meter" have not been realized, nor have 1960s and 1970s predictions of orbiting hotels, Lunar settlements, and human missions to Mars by the year 2001 (or sooner). Predicting the future is generally an uncertain endeavor and even the best crystal balls may be clouded by unknown factors. Among the challenges of predicting the future are understanding which concerns will be relevant, determining the reactions to those concerns, and identifying the critical decisions that will be made.[64]

Visions of the future, in contrast, present possibilities of what *could* happen, but some visionaries have been amazingly prescient. In 1812, Oliver Evans envisioned railroads criss- crossing the U.S., transporting people and cargo with steam- powered locomotives, decades before the national railroad network became a reality. In 1927, Fritz Lang, in his film *Metropolis*, depicted a city of 2026 with elevated highways wending their way past high-rise buildings, offices with video conferencing, electronic stock-tickers, and other developments that today are commonplace, or even seem outdated.[65]

[xx] From Goddard's high school graduation oration, "On Taking Things for Granted," June 1904
http://www.nmspacemuseum.org/halloffame/detail.php?id=11

A vision that addresses key issues is more likely to be realized. In addition to energy and environmental issues, other critical issues in the 21st century relate to distribution of resources, minerals, and commodities, especially water and food. Population increases will exacerbate all of the other issues. International relations will also be affected and there will be societal repercussions. A vision that addresses several issues will be more beneficial and will more likely gain broad support than single-issue visions. Technological advances can help, and space activities can have an important role in such advances, though non-technical considerations may be more important than particular technologies. However, having a vision of the future we want—that addresses issues in an achievable manner—is key to focusing our efforts.

5

Visions for the 21st Century

"There are two primary choices in life; to accept conditions as they exist, or accept the responsibility for changing them." — Denis Waitley

IN THE OCTOBER 2007 ISSUE OF NATIONAL GEOGRAPHIC, containing articles about global warming and biofuels, and their significance for the 21st century, there is also an article on the space program and the 50th anniversary of the launch of *Sputnik 1*. That article succinctly summarizes the history, current state, and near future of space activities, both government-funded and commercial efforts. What is apparent from that article is the importance of a unifying vision for space activities, and the difficulty of having one. Following the *Columbia* accident in 2003, Admiral Hal Gehman noted that "the U.S. civilian space effort has moved forward for more than thirty years without a guiding vision."

Some say that a foreign challenge similar to the impetus for Apollo is needed now to stimulate further advances in human spaceflight, such as a mission to Mars. However, the Apollo program also illustrates a risk to that approach: with a narrowly focused vision based on such a challenge, the motivation ends when the challenge is met. Former adversaries in space activities have become partners and many countries are developing the capability to reach space. The opportunity now exists to build upon cooperative efforts as the basis for our activities in space. A vision based on cooperation for mutual benefit can broaden the motivation and extend the rationale in ways that sustain advancement. Focusing that vision to address the critical issues of society could provide long-term motivation. How can a vision of space activities address the primary issues of the 21st century?[66]

• • •

THOUGH THE SPACE STATION was expected to be completed in the mid-1990s, to me, in 1984 when I became involved with the project, it represented the beginning of the 21st century. As an orbiting laboratory, observation platform, satellite servicing facility, manufacturing plant, and way point for deep space missions, the Space Station would open the way to a future of increased space activities. Called "the next logical step," the motivations for the space station included advancing scientific research and developing commerce, by manufacturing products and servicing satellites. The means to live in space would be developed further, as well, as we gained experience operating the space station, increasing our ability to have human missions away from the Earth.

Developing life support systems was the work I wanted to do, for which I hoped to combine my interests in botany and engineering. Even while working in the Design Integration Branch, I was involved with developing the life support system for the Space Station, researching the life support technologies to determine how the different life support subsystems could be configured into integrated systems. The purpose was to compare alternative configurations and evaluate the system-level mass, volume, power consumption, and other integration needs, and the results were used by the engineers preparing preliminary designs of the space station life support system.

In 1986 Marshall was named lead center for the environmental control and life support system (ECLSS) for the Space Station and, on September 2, I rotated to the Life Support Branch, which was staffing up for the new task. I was assigned the carbon dioxide (CO_2) removal subsystem, which suited me perfectly. Removing CO_2 from the atmosphere is essential and in a space habitat—lacking the dilution capacity of a large volume—the CO_2 concentration can exceed safe limits within days, poisoning the crew unless it is actively removed. Plants perform this function on Earth, but the means to grow plants in a space habitat were not sufficiently advanced (though efforts were being made to develop that capability), so another method was needed. For the early U.S. space missions through Apollo, a chemical sorbent (lithium hydroxide or LiOH) was used quite effectively, but this is an expendable material that cannot easily be regenerated. For a short-duration mission up to about two weeks, LiOH has advantages over alternative methods, but the Space Station would operate continuously for many years, and the resupply costs would be prohibitive. For *Skylab*, the only long-duration space experience that the U.S. had at that time, a regenerable CO_2 removal device had been used with great success. The sorbent material for this device was a zeolite molecular sieve that removed excess humidity as well, venting CO_2 and water vapor to space. One option for the Space Station was a version of the *Skylab* molecular sieve device, modified to enable the moisture to be returned to the habitat to reduce the amount of water to be resupplied. The CO_2 would either be vented to space or, perhaps in the future, delivered to a device to recover the oxygen, further reducing resupply needs. There were other, newer alternative devices for performing this function that had potential advantages, if they could be developed in time. The task of my coworkers and I involved evaluating the performance of these devices for each life support subsystem. The Marshall test facility was modified to enable comparison testing under simulated Space Station conditions. With this approach, NASA could select the technologies that best met the requirements for the Space Station.[xxi]

This was the work I wanted to do, so after completing my three-month rotation I gave my presentation to the lab director and requested to remain with the life support group. Upon the approval of the Design Integration Branch chief, and Randy Humphries, chief of the Life Support Branch, I was officially transferred. To the people who worked for him, Randy was famous for his yellow sticky notes. On Monday morning, if we found one or more yellow notes stuck to our desks, or to a report or presentation left on our

---

[xxi] In addition to CO_2 removal, other life support functions are oxygen generation, CO_2 reduction, trace contaminant control, atmosphere control and supply, water processing, urine processing, and waste processing.

chairs, we knew Randy had been in over the weekend. We found them on many Mondays. The note would be an action item or question usually written in red ink and in his distinctive handwriting. After a while, we learned to decipher it.

Randy had worked for NASA since the 1960s and had been deeply involved in developing the life support system for *Skylab*, for which Marshall had also been responsible. He knew that technical excellence was the foundation for NASAs achievements, and he endeavored to foster that excellence in everyone who worked for him. For example, when presented with a new concept that was not described clearly, he would ask "what's the physics?," to find out whether the person pitching the idea actually understood the concept.[67]

I experienced this emphasis on technical excellence first-hand shortly after I started in the group, when Randy tested my engineering acumen with a question: "Will a heated rod cool more quickly in a horizontal or vertical orientation?" I thought carefully before answering, concerned that there might be a trick. "Horizontal." He seemed satisfied with my response and then asked, "Why?" This time I answered more quickly. "Because of boundary layer effects due to convection. The thermal boundary layer reduces the rate of heat transfer, and it is much thinner when the rod is horizontal." I passed the test.[xxii]

With an expected launch date by 1992, six years away, my coworkers and I eagerly dug into our tasks, excited to be working on such an important project. The loss of *Challenger* was a major setback, but we expected the Shuttles to return to flight with

---

[xxii] There could have been a "trick" in that question. The answer that I gave assumed that the heated rod was in a gaseous environment on the Earth, or on some other massive body, and that free convection was the primary mode of cooling. In a vacuum or the "microgravity" of space, the situation would be different. In a vacuum (in space or on the Earth), without a surrounding fluid, there would be no convection or conduction, and in microgravity, even with a surrounding fluid, convection would not occur without some motive force such as a fan. In these situations thermal radiation alone (in a vacuum) or radiation with conduction (in a fluid environment) would be the mechanisms for heat transfer and the orientation would not likely be significant.

minimal disruption. The long delay, however, greatly affected the schedule for the Space Station in ways that would become clear only over time. The 21st century would not start quite so soon after all.

THE CHALLENGES OF THE 21ST CENTURY are unlike those of any previous century, in magnitude if not in substance. Supplying energy and acquiring material resources in environmentally responsible ways, while avoiding the worst aspects of climate change, are monumental challenges. Technology alone, advanced or not, will be insufficient to meet the challenges. What is required is understanding of the nature of the problems, knowledge of the possible solutions, creativity in applying solutions, and the desire to find sustainable, equitable solutions. As global challenges, cooperation and coordination of efforts on an international level are key to addressing them and will help ensure that appropriate technologies are developed—whether incremental improvements to existing technologies or innovative breakthroughs—and that they are effectively applied.

The increase in CO_2 emissions and their contribution to global warming must be broadly addressed. As the proposals mentioned in Chapter 1 indicate and as Bill McKibben writes in *National Geographic*, "The scale of the problem means we'll need many strategies." It will be expensive to implement these strategies, though the cost depends upon the methods chosen, but the cost of not addressing CO_2 emissions will be even greater.[68]

According to the International Energy Agency (IEA), a 50 percent reduction in CO_2 emissions by 2050 would cost about $45 trillion. The Group of Eight[xxiii] and others aim for 50 to 80 percent reduction in greenhouse gas emissions by 2050. Though claimed by some to be excessive, this goal actually may be inadequate. Even if greenhouse gas emissions were reduced 100 percent immediately, the Earth will continue warming and sea levels rising

---

[xxiii] Abbreviated the G-8, consisting of the major industrialized countries: Canada, France, Germany, Italy, Japan, the Russian Federation, the United Kingdom, and the United States.

due to effects of the additional CO_2 and other greenhouse gases already in the atmosphere. If, as others claim, it is necessary to reduce the CO_2 concentration in the atmosphere—currently 389 parts per million (ppm)—back to 350 ppm to minimize the effects of global warming, then we actually need to remove CO_2. Creative solutions will be needed, especially to achieve such negative CO_2 emissions.[69]

To increase the likelihood of finding solutions, in their book *Break Through: From the Death of Environmentalism to the Politics of Possibility*, Ted Nordhaus and Michael Shellenberger suggest increasing public investment in clean energy to $15 to $30 billion/yr. As an example of the effect this would have on creating new industries they describe how, in the 1960s, the Department of Defense "effectively guaranteed the market for microchips, allowing firms to grow and eventually stand on their own," becoming multi-billion dollar corporations producing microchips for a wide range of products that are ubiquitous today. Nordhaus and Shellenberger also emphasize the importance of education, to ensure sufficient clean energy specialists for the future.[70]

In *Cradle to Cradle: Remaking the Way We Make Things*, William McDonough and Michael Braungart promote the idea of designing products for reuse and eliminating waste altogether. According to them "pollution is a sign of design failure." In *Plan B 3.0: Mobilizing to Save Civilization*, Lester Brown advocates addressing not only environmental and energy issues, but also eradicating poverty and designing livable cities. Issues associated with energy and resource use, the environment, and quality of life are intimately linked, and the solutions that are most effective will acknowledge that linkage. [71]

THE UNITED NATIONS ADDRESSES MAJOR ISSUES in the world and the reality for many people is that the basic necessities are scarce. The UN promotes a vision for the future in which good healthcare, adequate food, and education are available to everyone. Cooperation among nations is a key part of this vision, which aims to ensure continued development while maintaining a sustainable environment. In 2000, the UN Millennium Development Goals were established to guide development efforts, which aim, by 2015, to:

> "1. Eradicate extreme poverty and hunger
> 2. Achieve universal primary education
> 3. Promote gender equality and empower women
> 4. Reduce child mortality
> 5. Improve maternal health
> 6. Combat HIV/AIDS, malaria, and other diseases
> 7. Ensure environmental sustainability
> 8. Develop a global partnership for development."[72]

These goals are bold, but progress is being made toward each of them. However, rising fuel and food costs, and the 2008-2010 economic recession threaten to reverse gains. Eliminating energy as an environmental or security concern and eliminating damage to the Earth's biosphere in our acquisition of resources would aid efforts to reach these goals.

DURING THE COURSE OF MY CAREER WITH NASA, I was involved with projects that were successful (*Spacelab* and *Hubble*), some that encountered serious obstacles until cancellation (the Space Exploration Initiative (SEI)), and, of course, the Space Station, that also encountered obstacles but was restructured and survived. Why do some of these projects succeed while others fail? Sometimes the reasons are technical (e.g., the National Aerospace Plane) and other times a result of mismanagement (e.g., the *X-33*) or loss of political or public support (e.g., SEI). But is there a more fundamental, underlying reason?[73]

In February 1988, President Reagan proposed a plan that included restoring the Space Shuttle to flight (following the loss of *Challenger*), completing the Space Station by 1997, "resuming the manned exploration of the moon," and "mounting a manned mission to Mars ... after the turn of the century." This vision was presented near the end of Reagan's second term as president. An editorial in the *Huntsville Times* stated, "Still, if Mr. Reagan was going to put forth a space agenda so late, it is comforting that his vision is so grand. Now we confront the task of building a constituency for this adventurous future." I shared in the excitement.[74]

But Reagan's plan assumed that programs such as Apollo succeeded *because of* their grandness and, so, he felt that a new grand vision was needed and that the constituency to support it would develop. That, however, is "putting the cart before the horse." The space program visions that have been proposed since Apollo certainly didn't fail due to limited grandness—a grand vision is simply insufficient to establish a project that will take years or decades to complete. Apollo succeeded because it addressed fundamental issues of the time. A better approach would be to devise a vision ("grand" or not) that addresses the issues we are currently facing.

According to John Marburger, "[t]he ultimate goal [of space activities] is not to impress others, or merely to explore our planetary system, but to use accessible space for the benefit of humankind... The idea is to begin preparing now for a future in which the material trapped in the Sun's vicinity is available for incorporation into our way of life..." Could doing this help to address critical issues of society? Could energy and environmental issues provide a unifying vision for space activities? Becoming a space-faring civilization and utilizing resources from space can be integral to the vision of our future, even helping to address the UN goals.[xxiv] By providing abundant energy and resources in ways that promote a sustainable environment and provide broad opportunities, our security also will be enhanced. That's a vision of what space activities could do to benefit all of us. The challenge is how to get to that future from where we are now.[75]

[xxiv] Acquisition of energy and resources from space would not occur within the timeframe of the UN goals, however, advances in technology and cooperation could occur that would support those goals.

6

New Challenges

"Challenges are what make life interesting; overcoming them is what makes life meaningful." — Joshua J. Marine

CHALLENGES CAN SPUR CREATIVITY in the effort to overcome them. To find solutions to the challenges, however, a vision of the end goal is necessary, as is the desire to reach for that goal. Especially for societal challenges, support in the form of institutional and financial backing is also required. Finding creative solutions to challenges often brings new capabilities and, therefore, new opportunities.

• • •

"TODAY WE DON'T HAVE A CRISIS. We have an opportunity." With these words, on July 20, 1989, while commemorating the 20th anniversary of the *Apollo 11* landing on the Moon, President George H. W. Bush announced a new project for NASA: the Space Exploration Initiative. Indications of the new program were "in the air" for months before President Bush made his announcement. Following the low point of the *Challenger* accident, the goals of the space program were reassessed, with the aim of developing a bold initiative for NASA. The result was the SEI with three major goals: to ensure that the U.S. led the exploration and development of space; to advance science, technology, and enterprise; and to build the infrastructure to support settlements on the Moon and Mars. The vision of this project was to send people back to the Moon to establish a permanent presence and then on to Mars by 2019. As presented in a report on the program, the rationale was "the human urge to expand the frontiers of knowledge and understanding and the frontiers where humans live and work." Accomplishing this vision would challenge our capabilities and would require advances in many areas, one critical area being the life support system.[76]

As the design for Space Station *Freedom* was refined, specific technologies were selected to perform the various functions, including the life support system. Alternative technologies that would not be developed for *Freedom* might be advantageous for later use, either for an expanded *Freedom* or for other missions, and were being pursued through several technology development programs. Developing these technologies would also support the SEI goals. In May 1989, I was reassigned to work on the advanced life support projects, covering the range of life support needs for these future missions, with emphasis on instrumentation for monitoring and control, for which Marshall was the lead center.[77]

The technical requirements for missions to the Moon or Mars exceed those for *Freedom*, driving the need for technological advances. Because of the long duration of the missions and the distance from Earth, the reliability and durability must be greater, equipment must be repairable or replaceable during a mission, and instrumentation must be self-calibrating when possible. The expense of launching mass to orbit emphasizes the need for efficiency and minimal size for equipment, and also drives the need to reduce the amount of resupply mass that must be lifted to space, by recycling water and oxygen, for example. When no available technology could fulfill the mission requirements, a new technology would need to be developed.

THE CREATIVITY OF ENGINEERS AND INVENTORS can be stimulated by posing a problem to them and requesting ideas on how to address it. One way that NASA (and other U.S. government agencies) does this is through the Small Business Innovation Research (SBIR) program. This program seeks new technologies to improve current capabilities or to provide new capabilities. Specific needs of NASA programs are published in the annual SBIR Solicitation, and, in a quite far-ranging appeal, any small company meeting basic requirements (e.g., an American company with no more than 500 employees) can submit proposals of innovative concepts to address these needs. The ideas proposed have some technical risk since they are to "push the envelope," but they are potentially high-payoff advances that can be developed into a useable technology within three to five years.[78]

"SBIR Subtopic Manager for the ECLS group" was added to my responsibilities in June 1989. After screening proposals to eliminate those that did not meet the SBIR guidelines or address the need for which they were proposed, two engineers, and sometimes three, reviewed each proposal to determine whether the proposers understood "the physics" of their concept and evaluate the likelihood of successful completion. We then ranked the proposals according to quality. Usually there were more high quality proposals than could be funded with the money available. Of the ones that were funded, those shown to be feasible during Phase 1 were recommended for further development with Phase 2 awards that, hopefully, would result in a prototype device that could be tested by NASA. Further development required alternative sources of funding for government use, and commercial investment to develop specific products for non-government use.

For a habitat in space, trace contaminants pose special challenges. Unlike on Earth, you can't just open a window and "air out" a space habitat—except in the most extreme situation—due to loss of needed gases that are costly to replace. So, removing trace contaminants is a critical function of the life support system and must be performed continuously to maintain sufficiently low concentrations in the air. Typically this is accomplished by a combination of catalysts or activated carbon that can be regenerated and non-regenerable sorbent materials that need to be replaced periodically. This approach works well for short-duration missions, but for longer missions, it would be better to use completely regenerable methods or, at least, to minimize the amount of expendable materials. In 1993, NASA was seeking an improved method to remove contaminants that did not require an expendable material. Precision Combustion, Inc. of New Haven, CT responded with a proposal to develop a technology they had initially conceived for automotive use. With further development, this approach might also be suitable for space habitats, greatly reducing the quantity of sorbent materials needed, and thereby reducing the resupply mass, as well as maintenance tasks for the crew. The feasibility of Precision Combustion's Microlith® trace contaminant removal concept was demonstrated during the Phase 1 contract and a functioning prototype was built during Phase 2, which was tested to verify the performance and suitability for spacecraft use. Over several years, the technology was developed to a maturity level of 6 (on an 8-level scale), at a cost to NASA of about $1,000,000 and was then ready to be used in a flight program. (This is a bargain for developing such a technology from concept to useable product.) The additional development also improved the device for automotive application. But the technology has broader commercial potential, such as for chemical and biological cleanup and for reforming natural gas for fuel cells. Automotive companies are also interested in the technology for generating power from paint booth waste. Another commercial application is with gas turbines, to improve performance and reduce emissions at electrical power plants. This technology has spinoff applications that could help both electrical power generation and transportation improve efficiency and reduce contamination of the environment.[79]

THE "PATHFINDER" PROGRAM BECAME "Exploration Technologies" in September 1989, with a goal of returning people to the Moon in 2001 and landing people on Mars by 2015. This was an ambitious schedule, especially due to the technical challenges for the new programs that would require development of new techniques to monitor, control, and provide the life support functions. To bring together the people and ideas needed to develop the technologies a series of meetings, workshops, and conferences were held across the country. NASA engineers presented the mission scenarios to researchers and contactors to clarify the requirements and identify the remaining needs. Promising technologies that were in the early stages of development, but had the potential to meet the requirements, were evaluated regarding their capabilities and the amount of additional development needed. [80]

The need for fresh food for these long-duration missions led to renewed interest in using plants, which would also refresh the air and water. No plant-based life support system had ever been used on board spacecraft, though some plant growth experiments were flown on *Skylab*, *Spacelab*, and on the USSR's space stations. Such a plant-based approach is called a "controlled ecological life support system" (CELSS), and would require developing plant growth facilities suitable for space habitats. With my interest in botany as well as engineering, I was excited about the possibilities.

Beginning in the 1960s, experiments were conducted in the U.S. and USSR using algae for air purification and as food. By the 1970s, experiments had progressed to include higher plants and people in the growth chamber. At the *Bios-3* facility in Krasnoyarsk, Siberia, completed in 1972, three people were sealed in a 315 m^3 (11,123 ft^3) chamber for up to 180 days at a time. These experiments verified the basic concepts, but also revealed that incompatibilities could lead to chemical imbalances and extinction of some species.[81]

In the late 1980s, Edward Bass funded the construction of *Biosphere 2* in Oracle, Arizona, to study biosphere dynamics. *Biosphere 2* is a much larger and more complex closed ecosystem than *Bios-3*, covering 12,800 m^2 (about 3 acres) and replicating seven distinct biomes[xxv] found on the Earth (Biosphere 1). For two years, from September 26, 1991 to September 26, 1993, eight "biospherians" lived in the facility, relying on it for air, water, and food. [xxvi] [82]

Complex systems often experience unexpected interactions, and that was certainly the case for *Biosphere 2*. The oxygen level decreased at a rate of 0.3 percent/month, until it stabilized at a level comparable to an elevation of about 1,200 m (4,000 feet). On two occasions supplemental oxygen was added. The oxygen loss is thought to be related to CO_2 absorption by the concrete base of the structure as it cured, removing CO_2 before the oxygen could be

---

[xxv] A biome is a geographical area with the same climate throughout and with similar communities of plants, animals, and microorganisms. Biosphere 2 had a rainforest, an ocean complete with coral reef, mangrove wetlands, a savannah grassland, a fog desert, an agricultural area, and a human habitat.

[xxvi] Eighty percent of the food consumed was produced in Biosphere 2.

separated by plant photosynthesis.[xxvii] Most of the vertebrate species and all of the pollinating insects died. But *Biosphere 2* demonstrated "the resiliency and plasticity of an ecosystem" and can serve as a precursor to closed biospheres for settlements in space, as well as provide insights about "Biosphere 1."[83]

As *Biosphere 2* was nearing completion, around the same time as the SEI program was underway, interest in self-contained habitats was increasing. Constructed habitats that provide a healthy, comfortable environment all share certain characteristics even when located in settings as different as the ocean floor and outer space. Sharing the results of development efforts and ongoing research could be beneficial to all biosphere or closed habitat projects. In the autumn of 1990, a group of engineers and scientists involved with closed habitat development met in Huntsville to discuss how such exchanges could be accomplished. One way was to form a non-profit organization to provide a forum for presenting the results of active work and, so, the Institute for Advanced Studies in Life Support was formed on October 25, 1990. Over succeeding months, plans were made to hold biennial conferences and to publish a journal, *The Journal of Life Support and Biosphere Science* (later renamed *Habitation*). The first International Conference of Life Support and Biosphere Science was held in February 1992 at the University of Alabama in Huntsville, featuring the results of the *BIOS-3* experiment, progress reports on the *Biosphere 2* experiment, and NASA and university research. As William Crump, the first editor-in-chief of the journal described the purpose, "Each time we were successful in getting people from disparate backgrounds together, we learned something new. When I understood the history of the *Biosphere 2* project … it really wasn't so different from the process of designing a space station. An open mind and a safe environment to share ideas were all that was needed." The journal and conferences

---

[xxvii] During the manufacture of concrete CO_2 is released, but as concrete cures it absorbs CO_2.
http://www.sustainableconcrete.org.nz/default.asp?pid=778&sec=738

are important venues for sharing the results of biosphere research and development, and for stimulating new advances.[84]

AFTER THE LAST *BIOSPHERE 2* MISSION ENDED in 1994, Columbia University took over operation of the facility to perform research on the effects of increased CO_2 levels on a complex biosphere. Higher atmospheric CO_2 levels were found to have several effects, some expected and some surprising. The higher CO_2 led to increased acidity of the ocean biome that, among other direct effects, contributes to coral reef death, due to sensitivities of coral to pH. The relationship between CO_2 levels and plant growth was found to not necessarily be a direct correlation, however, such that the more CO_2 in the atmosphere the more plant growth occurs. It isn't quite that simple. While some species follow that relationship over a considerable range, others will do so only to a more limited point, beyond which increased CO_2 provides no corresponding increase in growth. Among the effects of increased CO_2 levels, it was revealed that at the cellular level the number of mitochondria and the structure of the chloroplasts were changed, possibly reflecting a major shift in plant metabolism. These results show that in a complex system such as the Earth there are important interactions that are still unknown to us and effects do not always happen as we expect based on our limited knowledge. What may be thought to be an insignificant part of a complex system may turn out to be vitally important. Such critical influences of seemingly insignificant factors have been referred to as "butterfly effects" with results that seem to show randomness in complex systems. We have much yet to learn about complex systems in general, and the Earth's ecosystems in particular, to avoid undesirable consequences of our actions.[85]

The study of ecosystems and the interactions between the various parts is vital to understanding how our actions affect the environment, as well as to designing habitats in space. On Earth, complex interactions between the land, water, air, and solar radiation shaped how life evolved. Living organisms themselves became another complex factor in the overall system that is the Earth. When separated from the Earth's natural ecosystem, the essential functions must be reliably duplicated in order to maintain a healthy and comfortable environment. Oxygen and potable water must be provided, carbon dioxide and trace contaminants must be removed (as well as biological and solid waste), and the temperature and humidity must be controlled. All of these functions must also be integrated into a viable system. This requires considering how these functions interrelate and how synergistic benefits can be attained. Such considerations are the field of systems engineering, which requires broad technical knowledge and interdisciplinary management skills. To optimize systems, interconnections, cause and effect relationships, and—especially as complexity increases—the possible butterfly effects must all be considered when engineering systems. Sufficient knowledge and understanding is needed, as well as receptivity to the new ideas that may be key to successfully addressing the challenges.[86]

7

New Ideas

"Problems cannot be solved by the same level of thinking that created them." — Albert Einstein

"Man's mind once stretched by a new idea, never regains its original dimension." — Oliver Wendell Holmes

THERE ARE TWO GENERAL APPROACHES to dealing with challenges, that may be described as linear and innovative. The linear approach is to extrapolate current trends with incremental changes, while the innovative approach is to find alternate ways, perhaps even questioning the basis of current methods.[xxviii] Each approach has merit given appropriate circumstances and each approach can lead to disaster when applied inappropriately or without considering undesirable consequences.

Either approach may involve the development of new ideas, but the search for breakthrough technologies is an example of the innovative approach. Finding such breakthroughs has been key for successfully addressing challenges, but an overemphasis on finding a breakthrough can lead to missing creative ways of applying existing methods that may satisfactorily address a challenge. Even incremental changes, though, may require thinking "outside the box," and not being constrained by past limitations.

New ideas provide new capabilities, and thus the means for new achievements. Such advances occur as a sort of dance

---

[xxviii] What I refer to as the linear approach may also be called "incremental innovation" while the innovative approach may be called "disruptive innovation." For more detailed descriptions, see:
Laporte, Amaury, "Commercializing Disruption & the Power of Passion In the Emerging Fuel-cell Industry," Dissertation for Master of Arts in Strategic Management, University of Nottingham, Nottingham, United Kingdom, 2001, http://www.alaporte.net/diss/DISSERTATION-Amaury_Laporte.pdf

between insight and application, which spawns new insights and
further applications, in an ongoing creative interaction.

•••

AN ALIEN LANDS ON EARTH and approaches the nearest person
he sees, who happens to be an engineer. The alien explains
that he has a device that will reverse global warming, provide
virtually unlimited energy, and bestow abundant resources, all
without damaging natural ecosystems. He wants to present this
device to humanity and hands it to the engineer, then flies away.
The engineer considers whether to give the device to the president
of the U.S. or the secretary general of the UN, or to sell it on eBay.
After a moment of indecision, the engineer takes the device home
and ... disassembles it!

Engineers are a curious lot. They are curious about how things
work and will eagerly take a device apart to find out how it
operates. They are also curious about finding better ways to do
things, so the ending of the joke could be that the engineer figures
out how the device functions and builds an improved version.

At its best, engineering is a very creative endeavor. Modern
conveniences that we take for granted, from cell phones and MP3
players to air conditioners and magnetic resonance imaging (MRI)
machines, were all created by engineers. Applying the known laws
of physics in novel ways, engineers produce new capabilities that
enrich our lives. Functionality is generally the goal, more so than
beauty or style, but some engineered structures, such as the
Brooklyn Bridge in New York City, are widely considered to have
an artistic beauty that transcends their functional purposes.

The Brooklyn Bridge is also an example of how a seemingly
simple technological advance can have profound impacts. In the
1830s, Wilhelm Albert in Germany developed "wire rope" or
cables as an improvement to fiber rope and metal chains. In the
1840s, John Roebling, a German-American engineer, improved the
manufacture of cables and then designed suspension bridges using
them, including the Brooklyn Bridge, which, with a main span of
485 m (1,595 feet) and a tower height of 83 m (273 feet), was the

largest bridge in the world when it was completed, in 1883. Steel cables allow bridges to be built across expanses of water previously thought impossible to bridge, and the longest suspension bridge today (the Akashi-Kaikyo bridge in Japan) has a main span of 1,991 m (6,529 ft).[87]

Addressing our energy and environmental issues, new ideas are needed to improve energy efficiency of common products and to develop environmentally sustainable sources of energy. The common light bulb is an example of a product where improved efficiency can have significant benefits, because of its wide usage. Ideally a light bulb would convert 100 percent of the electricity it uses into light, shine with the intensity and quality needed, never burn out, not use hazardous materials, and be inexpensive. Except for the low conversion efficiency and limited life, the typical incandescent bulb comes close, but at a time when energy and material use are of increasing concern those exceptions drive the need for a replacement. Engineers are continuing to improve the incandescent light bulb and General Electric, the company founded by Edison, is developing incandescent bulbs using new materials that make them twice as efficient as current bulbs. Their goal is an incandescent bulb that is four times as efficient, comparable to compact fluorescent bulbs.[88]

Widely promoted as superior to incandescent bulbs, compact fluorescent bulbs convert much more of the electricity they use into light, thereby reducing the demand for electricity, and last longer than incandescent bulbs, reducing material usage. But they also contain a small amount of mercury and so must be disposed of carefully to avoid contaminating the environment, and they still have a limited life. Another approach, solid-state lighting (SSL), comes closer to the ideal than either incandescent or fluorescent lights. One type of SSL is especially promising, the light emitting diode (LED). LED lights are being developed for growing plants in space habitats, to efficiently provide light with the appropriate quality and intensity for optimum growth, but they are already ubiquitous for a growing list of Earthbound uses—as colored indicator lights on computers, cell phones, traffic signals, and other devices, as well as decorative lights, and they are used in some televisions. The Times Square New Year's Eve ball that welcomed in 2008, the centennial of that event, was lighted by 9,576 LEDs, and was brighter than the previous year, yet used less electricity. Development of white light LEDs has led to flashlights with 50,000 hour, or better, lifetimes. Wal-Mart is replacing the fluorescent lights in refrigerated display cases with LED lights that are expected to net up to 66 percent energy savings, also due to occupancy sensors and dimming capabilities. More recently, engineers have developed LED lamps that can replace screw-in incandescent or fluorescent bulbs. Efforts to improve the luminous efficacy (light output per energy consumed or lumens/W) and color quality are underway, and the Department of Energy is funding research with a goal of producing a market-ready, 160 lumen/W bulb by 2025. Such light bulbs will approach the ideal. Edison would be amazed![89]

We can also consider the characteristics of an ideal power plant. This is not as easy to describe as the ideal light bulb, but some general characteristics can be identified: the ideal power plant safely and inexpensively provides energy in the form needed, when needed, with 100 percent conversion from a source that is abundant, without producing hazardous byproducts or damaging natural ecosystems. How close can we come to this ideal?

Today, most electricity is generated by coal-fired power plants, but these are the antithesis of an ideal plant, producing toxic coal ash as well as releasing CO_2 into the atmosphere and involving the destruction of ecosystems during mining of coal, while having perhaps a 35 percent conversion rate of energy from coal to electricity. Coal-generated electricity is cheap only by ignoring environmental and other indirect costs. In comparison, natural gas-generated electricity is cleaner than coal, but still releases considerable CO_2. Nuclear power plants are somewhat better, at least they do not release CO_2 during power generation, although we still do not have a satisfactory way to dispose of the radioactive waste, which results in security concerns as well as health issues. Hydroelectric power generation does not release CO_2 either, but natural ecosystems are damaged by the creation of reservoirs and most rivers suitable for major power generation have already been dammed. They also are subject to changes in rainfall patterns, especially droughts.

Both solar- and wind-generated electricity avoid the worst environmental issues of coal, gas, nuclear, and hydro power, but they have limits, as well as environmental issues of their own. The first limitation usually mentioned is intermittency, though this can be addressed in several ways. Solar photovoltaic (PV) facilities provide power during the day when demand is usually highest, but they can also be used in conjunction with energy storage methods (including existing hydroelectric) to ensure availability at night, and solar thermal power plants can be designed to store enough heat to provide power through the night. Wind power "farms," when properly sited, have little down time and when used in conjunction with other methods such as hydroelectric can provide steady power.[90]

Environmental issues include the toxic materials used during manufacture of PV panels and the area needed to produce significant amounts of electricity. New manufacturing techniques such as thin film or organic semiconductors may allow PV panels to be made at much lower cost and without the toxic waste, but relying totally on solar energy would require covering vast areas with PV panels (or mirrors, for solar thermal electricity generation). Where rooftops can be used, that is fine—and much of the electricity used in the U.S could be generated by solar PV on the rooftops of commercial buildings—but deserts are also suggested, though it is less well understood what the environmental impacts would be. For one Mojave Desert solar thermal plant, for example, 800 hectares (2,000 acres) will be covered with mirrors to generate 250 MW of electricity. Considering that electricity use worldwide is in the billions of megawatts, an enormous area would need to be covered.[91]

For wind turbines, bird kills and habitat disturbance are environmental issues, but with careful siting and improved design, these impacts can be minimized. Other issues are that locations with suitably strong, steady winds are limited and are often far from electricity consumers, requiring the construction of transmission lines with resulting energy losses during transmission and associated habitat disturbance. According to a 2004 report from the National Academy of Sciences, there also may be limits to wind power due to "nonnegligible climatic change at continental scales," though far less so than with fossil-fueled plants. Wind turbines could safely provide for much more electricity, perhaps 20 percent or more of the total amount generated, compared to the less than 1 percent at present.[92]

Geothermal energy use is also increasing and the U.S. is the world leader with 2,936 MW capacity in operation. Geothermal power plants operate continuously and have a low overall environmental impact, but they also require water—that may already be in short supply—either to be in the ground or available to inject into the ground, and in some locations efforts to build geothermal plants have led to earthquake activity. Even so, geothermal energy could likely provide 10 percent, or more, of our electricity needs.[93]

Nuclear fusion power generation is one of the breakthroughs being pursued and is touted as the energy source of the future. Efforts to harness fusion energy go back to the 1950s, but progress is being made. In 2007, an international consortium of countries selected Cadareche, France as the site for the ITER, a 10 billion euro experimental reactor to test the basic concepts for controlled fusion and to verify the ability to generate more power than is needed to start the reaction and keep it going. To do so, the ITER must heat a plasma of deuterium and tritium to 10 times the temperature in the core of the sun. The goal is to finish construction and begin operation of the reactor by 2016, with a possible start date for a commercial reactor by 2030 (though these dates may be optimistic). Success, when it comes, will be a remarkable achievement, and will provide an abundant source of energy with minimal environmental impacts, but major use of nuclear fusion is several decades away, at best. None of these options is completely ideal, so on which one or ones should we focus our efforts?[94]

WHEN PRESENTED WITH ALTERNATE SOLUTIONS for a given challenge, how should we choose among the alternatives? Even if "the best" solution cannot be determined, it would be good to select a "better" solution. How can that be done? This situation was faced during development of the life support system for the space station. In the mid-1980s, when I first began working in the ECLS group at Marshall, the devices to perform the life support functions on board the Space Station had not yet been selected. There were multiple technologies capable of performing each function, championed by the companies that provided the devices being considered. The task of the ECLS group was to test and evaluate the capabilities of each so that selections could be made. The process involved testing at several levels, beginning with "stand alone" testing to establish baseline performance and to optimize the operation of each device, and then increasing levels of integrated testing with related devices interconnected to determine how well they functioned together and to identify any anomalies. Finally we conducted "human-in-the- loop" testing, which meant, for example, collecting metabolic wastewater, including sweat and urine, for testing the water processors and then having volunteers taste the purified water.[xxix]

When making selections among two or more alternatives, trade studies are performed that may use two general methods. The "advantage/disadvantage" method applies when characteristics are not well quantified, whereas the "weighted factors" method applies when characteristics can be quantified. For either method, the critical characteristics must first be determined. For the advantage/disadvantage method, the alternatives are rated relative to each other and, for the weighted factors method, numerical ratings are determined and multiplied by the weighting factor assigned to each criterion to obtain an overall score. The weighted factors method is generally preferred, but at times it may be necessary to make a selection based on the advantage/disadvantage

---

[xxix] Other types of testing include life testing to determine the durability of the devices and flight testing to ensure that any gravity-sensitive processes have been properly dealt with. At any stage of testing, unacceptable results may eliminate a device from further consideration or, less drastically, show where modifications are needed to improve the performance.

method, at least for some characteristics. Though it may not definitively select the "best" option, the advantage/disadvantage method can at least select a "better" option. Even with the weighted factors method the factors may be subjective or mission-dependent. For example, is mass more important than power consumption, equally important, or less important? If more or less important, by how much—twice, three times, one-and-a-quarter times, half as much? These are judgment calls, based on experience and understanding of the mission needs and capabilities. Greater power consumption may require increasing the mass of the power supply system. If solar arrays are used and limited power is available, then the weighting factor for power consumption would be high. In contrast, if fuel cells are used for the mission and they are sized for water production and produce excess power, the weighting factor would be low.[xxx] When more than one candidate technology meets the mission requirements with similar scores, then factors such as cost are used to make the final selection. The cost includes not just development or operation costs, but also costs related to logistics and maintenance: the total lifecycle cost. Non-technical factors may also be important in making the final selection, possibly overriding the technical selection.

CONSIDERING THE OPTIONS FOR POWER GENERATION, the "power plant" that is closest to the ideal is reduced consumption, either by improved efficiency or changes in standard practices. The DOE Energy Star program, net-zero energy building construction, and changing personal habits to reduce energy use are examples of this approach. Of the common electricity generation methods, each has serious shortcomings compared to the ideal power plant, and the renewable sources have their own limitations. Because of this situation, we need to consider novel methods.[95]

[xxx] Generally, fuel cells are sized for power production and produce excess water, so, it would be a matter of how much of an increase in power supply size would result. Adding photovoltaic capability may be easier than increasing the capacity of the H_2 and O_2 supply for fuel cells.

Other potential energy sources that have been proposed include
extracting methane from methane hydrates mined from the sea bed
(but this is also a fossil fuel and so would add to CO_2 concerns as
well as have issues related to mining the ocean); growing
hydrogen-producing algae or other microorganisms for energy
production; using sea turbines to extract energy from ocean
currents or tides; and converting radiation or low-level heat
directly to electricity. Even "cold fusion," first announced in the
1980s but then denounced, is being reconsidered and researchers
are still trying to determine whether "low temperature nuclear
reactions" actually occur. In 2009, in Norway, construction began
on an osmotic power plant, using the effects of osmosis between
salty and fresh water to drive a turbine. Another intriguing idea is
to loft wind turbines high into the sky, even to the jet stream,
where the winds are stronger and steadier. Such "flying windmills"
could provide continuous power, though they may need to be
reeled in during storms to avoid damage by lightning. None of
these methods is completely benign, however, since they all
interact with the environment and, when carried out on a large
scale, will likely have unintended environmental consequences.
Whether those consequences would be acceptable is not currently
known, but other methods may be able to avoid them completely.[96]

NEW IDEAS CAN BE CONCEIVED AT ANYTIME, while development of
those ideas often requires certain conditions. This was pointed out
by Buckminster Fuller when he visited Louisville in 1972. I was
about to graduate from high school and I went to hear him speak.
The message that Fuller presented was very exciting. He was
optimistic about the future and about our possibilities to increase
opportunities "to be a success." His mantra was "doing more with
less," and he pointed out that advances such as telecommunications
satellites—that provide more capabilities than transoceanic copper
cable of many times the mass—were leveraging knowledge to
accomplish tasks more efficiently. Fuller said that "through
improved materials and alternate systems ... we can produce ever
higher performance per each pound of material, minute of time,
and watt of energy ... (and) we are able to sustain all humanity at a

higher standard of living than heretofore experienced or dreamed of by any human." He also talked about the rate of technological advances and had a chart showing such advances over time. He observed that "there are ... shoulders, or plateaus, appearing on the chart. These shoulders are slowdowns when we have major wars— the American Revolution, various civil wars and World War I. They show that pure science does not prosper at the time of war— which is contrary to all popular notions. Scientists are made to apply science in wartime, rather than look for fundamental information."[97]

World War II is an example of warfare spurring technical advances—from radar to jet engines to strobe lights to rockets, the technologies developed for the war effort had broader benefits for society. The ideas that were developed into many of those advances, however, were actually conceived prior to the conflict. For example, liquid-fueled rockets were developed at a time of peace, but major development did not occur until World War II. The reason for the burst of advances during wartime is twofold: the receptiveness to new ideas and increased institutional support with funding. Military purposes simply provide the institutional backing and money to develop new ideas.[98]

But wars are extremely expensive.[xxxi] The technological benefits of wars can be achieved for a fraction of the cost, provided there is receptiveness to new ideas and the institutional backing to develop them. Promoting the development of new ideas without the manic urgency of war can produce even greater benefits for society. Edison's inventions were not motivated by military purposes, but the advances from just this one prolific inventor profoundly changed society and the way that virtually everyone lives.[xxxii]

[xxxi] As Benjamin Franklin wrote, in 1783, in a letter to Joseph Banks: "... In my opinion, there never was a good War, or a bad Peace. What vast additions to the Conveniences and Comforts of Living might Mankind have acquired, if the Money spent in Wars had been employed in Works of public utility!"
[xxxii] Thomas Edison said "I am proud of the fact that I never invented weapons to kill."

The Apollo program is also an example of the advances that can be made when receptiveness to new ideas is combined with institutional support. The sense of urgency for Apollo was similar to that which occurs during war, but the goal was to beat the USSR by reaching the Moon, rather than by fighting their military. NASA provided the institutional support to pursue new ideas needed by the space program, but the resulting advances spanned many fields, from communications to medicine to computers to battery operated power tools to scientific research to weather forecasting and climate monitoring to photography and on and on. We are continuing to reap the benefits of embarking on the Apollo program. Receptiveness to new ideas is an integral part of NASA's culture, as indicated by the SBIR program and more recent programs such as the Centennial Challenges competitions.[99]

The lesson here is that we do not need to rely upon a state of war to obtain great advances in creativity and innovation. This insight is important in making the best use of our available resources (money, time, research capabilities) to develop the new ideas needed to address the challenges we face.[100]

Challenging design environments, such as for space applications, drive technological advances that enable us to "do more with less," and use limited resources more efficiently. Even with advances in conservation and efficiency, the demand for electricity is expected to continue increasing in coming decades, enhanced by Jevons Paradox unless measures are taken to mitigate that effect.[xxxiii] Widespread use of electric vehicles, or fuel-cell vehicles if significant advances occur, will be a new, large

---

[xxxiii] In the mid-1800's, as the industrial revolution was reaching its stride, William Jevons realized that over time coal would become more expensive to mine as readily accessible coal was used up. He also realized that increasing efficiency would lower the relative costs of use, making it a more cost-effective fuel source which would tend to increase demand. The same effect relates to coal and oil use today. When a goal is to reduce consumption, or byproducts, to counter this paradox other actions must be implemented to provide for rational pricing, such as a CO_2 tax, to attain the desired results.
Jevons, William Stanley, "The Coal Question," 1865, http://www.econlib.org/library/YPDBooks/Jevons/jvnCQ.html

demand. The needs of the developing world to support a rising standard of living will also increase demand. In addition, as clean water becomes scarce in certain regions, the need to desalinate sea water or to purify other nonpotable water will be another increasing demand for electricity.[101]

Whether a technology is developed or not is often determined by non-technical reasons, including government policies. As James Hansen states Melvin Kranzberg's *First Law of Technology*, "No technology is absolutely, by-its-very-nature 'good.' And none is bad. But neither is technology ever *neutral*." And technical factors do not always take precedence in making decisions about technology. "Politics and culture ... often override good technical or engineering logic. *And they should*." When deciding which ideas and technologies to develop, it is important to consider not just whether they will work, but whether they will support the objectives of a vision.[102]

If fossil- and nuclear-fueled power plants are not acceptable in a sustainable, secure vision of the world; efficiency improvements and conservation are insufficient; and currently available renewable sources are too limited, we will need alternative large sources of electricity (and transportation fuels) that are consistent with the end goals, including reducing CO_2 levels in the atmosphere. Fortunately, we do not have to choose only one method. As with financial investments, diversity is the best approach. The CMI report, for example, suggests that multiple approaches are needed—planting trees to remove CO_2 from the atmosphere; increasing conservation and efficiency to reduce energy demand; increasing the use of renewable energy; and increasing the use of biofuels, especially those made from waste. These actions would all help in the near-term and, with continued improvements, may be sufficient to provide the electricity and transportation fuels we need in a sustainable manner. In the long run, though, a way to provide energy that does not involve even residual waste or harm to the environment would be better, perhaps including methods not currently considered or those yet to be developed.[103]

Creativity is essential in addressing the challenges we face. Both fundamental research and product development require creativity and both are needed for developing new technologies. Any advances in technology made to meet the performance and capability requirements of space activities could provide spinoffs that improve energy efficiency or reduce waste and byproducts on Earth. Techniques developed to analyze information, control processes, and optimize systems in space could benefit a broad range of Earth applications. But that is just the beginning. Ongoing research continues to advance all areas of space-related fields. New types of materials such as aerogels and carbon nanotubes have, potentially, enabling roles in future space activities. They could also improve the energy efficiency of machinery and appliances, and of the construction and day-to-day operation of buildings. Carbon nanotubes tailored to have greater conductivity could even replace copper wire, thereby reducing losses in our electrical distribution network. Advanced techniques for making thin-film solar cells could enable the manufacture of large solar arrays without the toxic byproducts. Each of these technological advances could also play enabling roles in the utilization of space to address Earth-bound issues.

8

The Promise of Space

"The only way to discover the limits of the possible is to go beyond them into the impossible." — Arthur C. Clarke

"It is humanity's destiny to explore the universe. When we start thinking and working on that cosmic level, we will transcend our parochial differences and tribal natures and become global creatures, solar system creatures. Then we will figure out where we fit in." — Story Musgrove, astronaut

"SPACE EXPLORATION OUGHT TO DO SOMETHING for our civilization," according to William K. Hartmann, in 1992. He went on to say space activities should "be driven not only by interesting intellectual challenges, like the origins of the solar system and of life itself, but also by practical concerns about what is happening on Earth: changes in our climate, impending exhaustion of mineral and fossil fuel reserves, and pollution by heavy industry. The program that we can choose to shape would be designed not only to give us science facts, but to see if we can demonstrate human capabilities in space; to gain knowledge about how climate changes work; to discover metals and other resources in asteroids and to utilize them; and to demonstrate the potential for utilizing solar energy in space. The program would be a blend of data gathering and exploration, research and adventure, robotic and human activity." Given setbacks that were still fresh in the public memory at the time (*Challenger*, *Hubble*, the USSR *Phobos 2* mission to Mars, etc.) that cast questions about space activities, Hartmann asks "Do we still have the will, funding, and ability to explore space?"[104]

• • •

I N 1968, ARTHUR C. CLARKE wrote *The Promise of Space*, in which he describes the space program up to that eve-of-landing-on-the-Moon year and projects what could happen through the end of the 20th century and beyond. He includes speculation on settlement of the solar system and interstellar travel and concludes with a discussion of the "means and ends," the motives for going into space. Clarke states, "With the expansion of the world's mental horizons may come one of the greatest outbursts of creative activity ever known," comparable to the Renaissance period of western history. "The crossing of space—even though only a handful of men take part in it—may do much to reduce the tensions of our age by turning men's minds outward and away from their tribal conflicts. It may well be that only by acquiring this new sense of boundless frontiers will the world break free from the ancient cycle of war and peace." He presented a bold vision of a new age.[105]

From the vantage point of 40 years later, the results are hopeful, but mixed. The Cold War tension faded following the end of the USSR, and Russia joined the partnership of the *ISS* and other space projects. But even as we enter the second decade of the 21st century, conflicts and wars are not yet relegated to history. Space projects, however, can enhance international relations. As Michael Griffin, former Administrator of NASA, said, "The most enlightened, yet least discussed, aspect of national security involves being the kind of nation and doing the kinds of things that inspire others to want to cooperate as allies and partners rather than to be adversaries. And in my opinion, this is NASA's greatest contribution to our Nation's future in the world." Projects that involve international cooperation provide "the perfect opportunity to bind [nations together] as partners in the pursuit of common dreams. And ... all will be less likely to pursue conflict." Such cooperation in mutually beneficial pursuits increases the security of the world.[106]

Currently, several countries have programs or plans to study the Moon with the aim of sending people there for long-term habitation. Rather than "go-it-alone" approaches, working together for a common purpose could prove to be more beneficial.

Realization of such a vision could be a step toward Clarke's "promise of space."

BUT WHY SHOULD WE EMBARK ON RISKY ENDEAVORS such as space activities? Greed, fear, and curiosity have been offered as reasons. These, certainly, are base motivations, and may be better explanations than "an innate urge to explore" that is sometimes given as a more noble rationale. Greed motivates those who seek wealth with minimal effort and for whom material wealth is the prime objective in life. The miners who went to California in 1849 or to Alaska in the 1890s seeking gold were often motivated by greed (as well as adventure), though few actually became wealthy. Direct comparisons with the gold rushes have been made for proposed space activities, but the current cost and difficulty of access to space limits greed as a motivation. In addition, only a small percentage of people are driven by greed—the number of '49ers was in the tens of thousands, not millions—so, even with low-cost access, greed is not conducive to broad-based action.[107]

Fear also can be a powerful motivator, especially for intense, focused action. Fear of the technological capabilities of the USSR was the initial motivation for the Apollo program and fear is the underlying motivation for those who claim that current foreign competition is a reason for space activities, as well as for those who seek to spur action to prevent asteroid impacts, a slight, though very real, possibility. As a motivator, fear requires a specific threat, real or imagined, so the motivation ends when the threat is neutralized. As a result, fear is not conducive to long-term action.[108]

Curiosity as a motivator can lead to impressive advances. Robotic missions to other planets and telescopes such as *Hubble* address scientific curiosity and have revealed remarkable aspects of the universe, enhancing our understanding. However, the level of funding for such scientific space missions has been rather modest throughout the space age, indicating a relatively low limit to the extent of support. This is possibly because curiosity motivates action to address specific questions: What is over there? How does this happen? etc., but it is usually limited in scope, with each question championed by only a small group. Curiosity, therefore, is usually not a primary motive for long-term or broad-based action.[109]

To obtain long-term and broad-based support, space activities must address primary issues of broad concern and widely distribute the benefits of those activities. There is another, more powerful, motivation that is consistent with these requirements and that may provide a reason for open-ended action. I have already mentioned it numerous times in this book. In a word, that motivation is *opportunity*. As a motivation, opportunity *is* conducive to long-term, broad-based action, specifically the opportunities to effectively address energy, resource, and environmental issues; the opportunities to enhance cooperation among nations; and the opportunities for all to have better lives. These opportunities are the promise of space.

The issues currently faced by society do not require settlement of space, though addressing those issues through space activities would develop the means for settlement. While explorers may be motivated by greed, fear, or curiosity (or some combination) they are not the people who are likely to become settlers. Throughout history, settlers were motivated by the opportunities to be found by settling a new frontier—the opportunity to improve their lives, to have a place of their own, to raise their families in dignity, and to ensure a more secure future. The same is likely to be true for settling space. Opportunities, however, can only arise when we have prepared for them. For settlers of the American West, once the factors that enabled them to do so were in place, they responded by the millions, settling throughout the region. How can we prepare for future utilization and settlement of space?

Part III

The Way to the Future

"A vision without a task is but a dream, a task without a vision is drudgery, a vision with a task is the hope of the world." — From a church in Sussex, England, c. 1730 (attributed)

"As for the future, your task is not to foresee it, but to enable it." — Antoine De Saint-Exupery

• • •

WHATEVER THE SPECIFICS OF A VISION OF THE FUTURE, to become reality the means to achieve it must be present. For space activities to be most useful for addressing earthbound issues, the means to get to space must be reliable and affordable. Once in space, the means to live and function there comfortably and remain healthy are needed. Due to the magnitude of the effort to utilize space, cooperation among diverse groups (countries, companies, organizations) will be necessary that may require new approaches to interaction.

The vision presented in this book is bringing the solar system into the realm of human activity to address issues on Earth. The motivation is the seriousness of those issues. However, the means to enable this vision are not yet fully developed. The task before us now is to develop the means.

9

The Means, Part I: Getting to Space

"Once you get to Earth orbit, you're halfway to anywhere in the solar system." — Robert A. Heinlein

REACHING SPACE IS A MAJOR HURDLE and doing so at a low enough cost that it is affordable for major activities in space is an even greater hurdle. The reason is that it takes an enormous amount of energy to accelerate to the 7.6 m/s (17,000 mph) or so needed to orbit the Earth just above the atmosphere, and currently the cost using rockets is thousands of dollars per kilogram or pound. Finding lower-cost ways to reach—and move around in—space will be crucial to fully utilize the opportunities there. An alternative to rockets is needed that provides the same leap in capability as the transcontinental railroad or the jet engine did for settlement of the American West and air travel, respectively.

• • •

AFTER THE FOURTH LANDING OF *COLUMBIA* on July 4, 1982—completing the test phase of the Space Shuttle program—President Reagan compared that mission to the completion of the Transcontinental Railroad in 1869. Expectations were high that the Space Shuttle would significantly reduce the cost of access to space, allowing greatly increased activity, with a planned launch rate of one every other week. But it did not turn out that way. To obtain the full benefits of space activities, it is imperative that we have an efficient, safe, and low-cost means to get there.[110]

Development of the means to accomplish major achievements generally begins decades before the achievements occur, thus, the research and development of the technological means of future space achievements is happening now. Private efforts and government-sponsored research are seeking ways to reduce the cost of access and improve the capabilities of spacecraft in a variety of ways. NASA has several programs to do this, from the New Millennium program demonstrating new technologies, such as the ion engine of *Deep Space 1*, to the Centennial Challenges program that offers cash prizes for development of specific technologies, such as improved space suit gloves. Several private companies are also developing low-cost access to space. Orbital Sciences Corporation can place a 454 kg (1,000 lb) payload into LEO for $20 million to $25 million with their *Pegasus* rocket. SpaceX is developing the *Falcon*, expected to carry a 569 kg (1,254 lb) payload into LEO for $6.7 million or less. The Defense Advanced Research Projects Agency (DARPA) wants to develop a launch vehicle able to carry a 454 kg (1,000 lb) payload into low Earth orbit (LEO) within 24 hours at a cost of no more than $5 million for the rocket and launch. For human access to space, Burt Rutan developed the *SpaceShipOne* suborbital vehicle—using an innovative feathering technique to ensure proper orientation of the vehicle when reentering the atmosphere—to win the Ansari X-prize and is now developing more advanced vehicles, intending to launch people into orbit. Others are planning similar capabilities, addressing the challenges in creative ways.[111]

The Limitations of Rockets

SEPTEMBER 11, 2001 WAS A TRAGIC DAY for the U.S. and for the world. The terrorist attack with hijacked airplanes killed over 3,000 people, most in New York. The shock and initial disbelief was reminiscent of the *Challenger* accident. So, too, was the inability to foresee the consequences. The attacks set off a chain of events that led to two wars, many more deaths, and an expensive course of action that resulted in the U.S. becoming increasingly indebted. It is hard to know whether that is the result that the hijackers wanted, but one can imagine that they would be pleased with the effects of their actions.

Following the attacks, air travel in the U.S. was halted for three days while security measures were implemented at airports across the country. The interruption provided an unprecedented opportunity for meteorologists to study the effects of air traffic on the environment. As jets fly through the cold air at high altitudes, they leave wispy trails that form thin clouds due to condensation of moisture around microscopic particles in the exhaust. During the three days without air traffic, these condensation trails (contrails) were not being made and meteorologists noticed a startling effect on temperature patterns. Contrails, as do natural cirrus clouds, reflect some of the incoming solar radiation to space in the daytime, and hold in thermal radiation from the ground at night. These are counteracting effects, and what was discovered is that without the contrails the daily variations in high and low temperatures increased by 1.1 degrees Celsius (2 degrees Fahrenheit). Since contrails cover only about 0.1 percent of the sky on average (though with regional concentrations as high as 20 percent) this shows a potent effect on temperatures and possibly indicates other significant environmental effects. Such atmospheric effects are not limited to jet airplanes.[112]

The fire and roar of the barely contained raw power of a rocket launch is a heart-pounding thrill to experience, even as a spectator, but when a Space Shuttle or other large rocket (whether Russian, European, Chinese, etc.) is launched, it, too, leaves a trail of exhaust in the sky. The exhaust from the Shuttle main engines is 97 percent water (those big white clouds spewing from the Orbiter). About half of that water ends up in the upper atmosphere, where, because of the extreme dryness at high altitudes, even the amount of water from one Shuttle launch has a noticeable effect. Thin clouds form that may cover thousands of square miles and have been found to travel as far as the north or south polar regions. Including the solid rocket booster exhaust, hydrogen chloride (HCl) and other exhaust products are released into the atmosphere during each Space Shuttle launch. Large clouds of hydroxyl ions (OH–) have also been reported. The effects of these exhaust products are not known to be harmful and for a small number of launches—such as the few dozen each year currently worldwide— the effects are minimal. The atmosphere can recover between launches, so they should not pose a long-term impact. Increased activities in space, however, would mean a greater number of rocket launches. If we increase the frequency so much that the atmosphere does not have time to recover between launches, the effects would be compounded, potentially causing harm which might limit the number of rockets that can be safely launched.[113]

In addition to the effect of the exhaust products on the atmosphere, the hydrogen (H_2) required to fuel the Shuttle is made from natural gas in a process that releases CO_2. Also, a complex launch and mission control facility is needed to monitor the vehicle performance and mission status, there are constraints on acceptable weather conditions that can cause significant launch delays that increase costs, and extensive, time-consuming, and expensive preparation and follow-up is required for each flight. For space activity to increase significantly, other means of access to space will be needed that do not have these constraints.[114]

New Ideas for Reaching Space

IT MAY BE POSSIBLE TO MAKE ROCKET FUEL that does not have environmental issues. In 2004, Mikhail Eremets and colleagues at the Max Planck Institute for Chemistry in Germany reported that they had made "polymeric" nitrogen, consisting of four nitrogen atoms. When this form of nitrogen transforms into inert molecular N_2 (the main component of air) a tremendous amount of energy is released. Such a fuel would be about as environmentally benign as possible and could eliminate concerns about rocket exhaust. But even better methods of reaching space may be feasible.[115]

The rockets currently used are versions of ballistic missiles intended to send bombs around the world. These rockets were designed to reach a target rapidly. To deliver people and cargo to space, however, this is not a requirement. If an alternative method will provide access without the constraints, it does not need to be so quick.

For decades, balloons have been released that rise into the upper atmosphere to the edge of space to collect air samples, monitor conditions, and carry telescopes to view space or experiments to study cosmic rays. One intriguing idea for reaching space without rockets, referred to as Airship-To-Orbit, would use a balloon-type vehicle that is two kilometers (one-and-a-quarter miles) long. This vehicle would first float to 60 km (37 mi or 195,000 ft) into the sky, from a platform at 30 to 42 km (100,000 to 140,000 ft) that is buoyed aloft by even larger helium-filled balloons. An electric propulsion engine (such as an ion engine) would then gradually propel the vehicle to orbit, taking 3 to 9 days to reach orbital velocity. There are significant challenges to this concept that have yet to be resolved, such as dealing with micrometeoroid impacts, gas loss, and building an ion engine powerful enough to overcome the drag such a large vehicle traveling at hypersonic velocities would be subjected to, even in the tenuous upper atmosphere. The environmental effects of operating such engines in the upper atmosphere would also have to be carefully monitored and evaluated. But if the airship-to-orbit idea is successfully developed, it could open the way for greatly increased space activity. However, there is another concept for reaching space with even greater potential, if a sufficiently strong material can be made into a cable or ribbon.[116]

A SPIDER WEB IS AN UNLIKELY STRUCTURE. From a high point, a spider extrudes a single fine strand of sticky silk. Even a gentle breeze will carry this strand until it touches something—a leaf, a twig, a wall—where it sticks. Once the initial strand is firmly attached, the spider gingerly walks across it, releasing a thicker, stronger strand of silk as it goes. This strand serves as the beginning of an intricate web. The silk consists of a polymerized protein that is about 0.15 microns (thousandths of a meter) in diameter, much finer than human hair with a typical diameter of 60 to 80 microns. Although finer than human hair, spider silk is stronger, much stronger. Where the strength under tension of hair is about 200 megaPascals (MPa) (29,000 pounds per square inch (psi)), for spider silk it is greater than 1,000 MPa (145,000 psi).[xxxiv] Spider silk is, in fact, stronger than many alloy steels and is among the strongest known natural fibers. A rope of spider silk about the

_____

[xxxiv] The strength is the maximum tensile force divided by the cross-sectional area.

diameter of a pencil could stop a jet airplane landing on an aircraft carrier.[117]

There are materials that are even stronger than spider silk. In the 1960s, Dupont developed Kevlar®, a form of nylon that has additional chemical groups and a structure that gives it greater strength and stability than more typical nylon. The tensile strength of Kevlar is about 3,600 MPa (525,000 psi), which is able to stop the penetration of bullets and shrapnel, making it useful for such things as body armor for soldiers and reinforcement for automobile and bicycle tires. Kevlar is amazingly strong, but even stronger materials are being developed.[118]

Carbon has characteristics that are not found in other elements on the periodic table. One characteristic is the ability to bond strongly to other carbon atoms to form long chains, as well as bond to hydrogen, oxygen, and other atoms. This makes carbon well-suited to make plastics (such as nylon and Kevlar), and proteins (such as spider silk). Carbon also has more solid or pure forms. Coal, too, is made of carbon—the remains of ancient life, as are diamonds—the hardest known natural substance. In 1985, chemists Richard Smalley, Robert Curl, and Harold Kroto, at Rice University in Texas, serendipitously discovered another type of pure carbon structure. While studying aggregates of carbon atoms and molecules ("clusters"), they found unexpectedly regular carbon structures—made of 60 carbon atoms joined in a sphere with a pattern of hexagons and pentagons. These carbon structures looked very much like a soccer ball or a geodesic dome designed by Buckminster Fuller, although they measure only in the millionths of a meter (nanometer). Because of the similarity with geodesic domes and soccer balls, the structures were named Buckminsterfullerene, or Buckyballs. In 1991, other shapes, including rings, tubes, and sheets, having the same basic arrangement of carbon atoms as Buckyballs, were discovered by Sumio Iijima in Japan. These were determined to be a new class of carbon compounds and the name Fullerenes was applied to the entire class. Because of their small size, they are called nanoparticles, with the tube-shaped ones called nanotubes. To appreciate just how small in diameter they are, consider that, according to Boris Yakobson and Smalley, "nanotubes sufficient to span the 250,000 miles between the earth [and] the moon at perigee[xxxv] could be loosely rolled into a ball the size of a poppyseed."[119]

Carbon nanotubes are also phenomenally strong, with a theoretical Young's Modulus of 630 gigaPascals (GPa)[xxxvi] (92,000,000 psi) and a tensile strength of about 130 GPa (19,000,000 psi). Researchers are just beginning to learn how to work with these microscopic particles and make them into useful forms while maintaining their exceptional strength. Manufacturing long fibers, for example, that have the strength of individual nanotubes is a challenge, but progress is being made. In 2000, Philippe Poulin in France developed a way to extrude fiber from carbon nanotubes suspended in a liquid. The fibers were flexible, but not very strong. Building on that work, in 2003, Ray Boughman, at the University of Texas in Dallas, developed a method of producing fibers as thin as a human hair but 17 times stronger than Kevlar. In collaboration with Ken Atkinson of Australia, they were able to develop a method to make nanotube ribbon. Such strong ribbon would provide new capabilities, including sheets capable of supporting 50,000 times their weight. A sheet made from only 280 gm of carbon nanotubes could cover a hectare (4 ounces/acre). Currently carbon nanotubes cost about $454/kg ($1,000/lb), but continued advances in production might reduce that to under $23/kg ($50/lb). Development of such a low-cost, high-strength material would open a range of new possibilities. Could such a material be used to provide access to space?[120]

[xxxv] Perigee is the closest distance an object orbiting the Earth comes to the Earth. Apogee is the farthest distance away such an orbiting object goes. When an orbit is circular these distances are the same.

[xxxvi] Young's Modulus (E) is also called the Modulus of Elasticity and is a measure of the stiffness of a material. It is a ratio of the tensile stress (the tensile force (M) divided by the cross-sectional area (A), Newtons/m^2 or pounds/in^2) in a material, divided by the tensile strain (the increase in length (l) divided by the original length (L), m/m or in/in). In equation form: $E = (M/A)/(l/L)$ For a cable, a stiffer material means less sag in the cable.

IN THE OCTOBER 1945 ISSUE OF *WirelessWorld*, Arthur C. Clarke proposed placing communications satellites above the Earth's equator at an altitude such that their orbital period would be synchronized with the Earth's rotation period—geosynchronous Earth orbit (GEO). Thus, from the Earth they would appear to be stationary. He pointed out that only three satellites, equally spaced around the Earth, could provide communication coverage of the entire world. Less than twenty years later, in 1965, *Early Bird*, the first communication satellite, was placed in such an orbit, and there are now over 300 satellites in GEO. In his 1979 book *The Fountains of Paradise,* Clarke describes another idea using GEO, construction of an elevator to space. This was an updated version of Tsiolkovsky's idea in 1895. In 1960, Yuri N. Artsutanov, a Russian engineer, had proposed the same idea, but it was published only as a non-technical story in a Sunday supplement to *Pravda*. In 1966, a group of American oceanographers led by John Isaacs published a short article in *Science* on using cables suspended from a satellite to launch "masses into orbit." Neither of these articles came to the attention of aerospace engineers. Then, in 1975, Jerome Pearson, working at the Air Force Research Laboratory, independently invented the space elevator and published a technical paper in *Acta Astronautica*. That is where Clarke first encountered the idea and he based his book on Pearson's work.[121]

For a cable with a length of 35,800 km (22,240 miles)[xxxvii], the distance to GEO, the cable would need to have a tensile strength of at least 63 GPa, or, taking the density of the cable into account, a minimum of 35 GPa-cubic centimeter/gram (cc/g) with 80 GPa-cc/g desired. To reduce the overall mass of the cable, it would be tapered to provide only the strength needed at any point along its length. The stronger the cable material, the smaller the taper ratio can be, significantly lightening the entire elevator and reducing the amount of initial mass that would need to be launched into orbit. Kevlar is not strong enough, but carbon nanotubes have a strength

---

[xxxvii] The actual length would be much longer due to the counterweight located beyond GEO, but the center of mass and the location of greatest load would be at or near the GEO position.

well above that required. The challenge now is to make cables that maintain this strength and are suitable for a space elevator.[122]

It may seem like an impossibly fantastic goal, and Clarke is famous for saying that a space elevator will be built 10 years after people stop laughing about it, but progress is being made and people are beginning to take the idea seriously.[xxxviii] The space elevator was the theme for the 2005 Clarke-Bradbury International Science Fiction competition and, since 2005, competitions have been held to promote development of the technologies needed for such an elevator. The Spaceward Foundation manages the competitions for NASA as part of its Centennial Challenges program, which offers a total of $4 million in prizes for a series of competitions over five years addressing the cable and the elevator car or climber. The cable strength competition has prizes of $900,000 and $1.1 million for attaining a breaking force of 1 ton and 1.5 tons, respectively, for a cable that is 2 m long and weighs no more than 2 g. The best performance so far has been 0.72 tons. For the climber competition, each year the height of the cable to be climbed and the required speed of the climber are increased and, for 2009, the challenge was for the climber to reach 1 km height at a speed of 2 m/s (for a $900,000 prize) to 5 m/s (for a $1.1 million prize). At the 2009 games, the climber by Team LaserMotive achieved an average speed of nearly 4 m/s, winning the $900,000 prize. The competitions are open to anyone and teams from the U.S. and several foreign countries enter. The hope is to award the top prizes by the 2010 competition. Much additional work will be needed to develop an operational space elevator, but this competition spurs progress.[123]

Once a sufficiently strong cable and fast climber are developed, how would a space elevator be constructed? Just as a spider begins a web with a single strand of silk that is reinforced with others to support the web, the Brooklyn Bridge was constructed by building up the cables wire by wire using a "traveler" to run the wires between the anchorage points on each side of the East River. A

[xxxviii] This was reduced from Clarke's original statement in 1979: "The Space Elevator will be built about 50 years after everyone stops laughing."

similar technique would be used to build a space elevator cable. The initial strand would be launched by a rocket to GEO, from where it would be unreeled until it touched the Earth near the equator and attached to a platform, most likely floating in the ocean. Climbers would then add ribbons to the initial one until the desired size and strength is attained. The first space elevator would likely cost $7 to $10 billion and take about 3 years to construct. But once the first one is operational, the cost of "launching" spacecraft would drop considerably and it would be easier and cheaper to construct additional elevators. A report by the NASA Institute for Advanced Concepts optimistically concludes that it may be possible to construct the first elevator as early as 2018. Climbers, traveling at about 322 km/h (200 mph), would reach GEO in about a week. As more elevators were constructed there would eventually be a web of elevators radiating out from the Earth's equator similar to the spokes on a wheel, providing greatly increased access to space without the environmental concerns and operational constraints of large rockets.[124]

Besides the cable strength and the climber, there are other technical challenges for a space elevator. The tendency for the cable to lean as a climber ascends (or descends) due to the gain (or loss) of angular momentum will need to be considered in the design, as will the natural frequency of the completed elevator system, to avoid excessive tension and stress on the cable. Protection from the Van Allen radiation belt through which the elevator would pass also will be needed for the passengers and the cable. After 50 years of rocket launches, there is considerable debris orbiting the Earth, from paint flakes to rocket casings to dead satellites. These would pose a threat to the cable, as would micro- and macro-meteoroids. A means of shifting the cable to avoid them or a way of clearing the debris would be essential, and the ability to maintain and repair the cables will also be needed.[125]

Long cables have already been tested in space. The Tethered Satellite System (TSS), a joint U.S./Italian experiment, was flown in 1996 to study the electrodynamics of a tether system in the ionosphere (the electrically charged upper layer of the Earth's atmosphere) and to gather information on how the thrusters on the

TSS interact with the ionosphere. The tether was 20 km (13 miles) long and made of a conductive form of Kevlar. One goal was to generate electricity and as the TSS was slowly reeled out electricity was generated in the tether. So much electricity that, when the TSS was extended almost the full distance, electrical arcing with the release boom in the Space Shuttle cargo bay caused the tether to break. Before it broke, however, the data collected indicated that currents generated were much greater than expected (about 0.5 ampere, three times what computer models predicted) and the voltage was as high as 3,500V. A longer cable would produce even more electricity.[126]

The results of the TSS may be relevant to a space elevator. Carbon nanotubes can carry electrical current at densities greater than copper wiring and nanowires have been produced that are superconductive (i.e., have zero electrical resistance) at a temperature less than 20 Kelvins (-424°F), with indications that superconductivity can occur at much higher temperatures. Although a cable to a GEO station would not generate significant electrical current—because it would be moving *with* the Earth's magnetic field rather than *through* it as the TSS was—there is another option for a space elevator that may make use of this effect. A space elevator does not have to be geosynchronous. A shorter elevator could orbit the Earth at a closer distance and would not require as strong a cable because of the shorter length. (Such an elevator could be an interim step toward a GEO elevator.) Since it *would* be moving relative to the Earth it could not be attached, but the end of the elevator could be just above the atmosphere. With a center of gravity at an altitude of, say, 4,000 km (2,500 miles), where the orbital period is about three hours, the end of the elevator would have the same orbital period and be moving at about 13,000 km/h (8,000 mph), less than one half the speed otherwise needed to orbit just above the Earth's atmosphere. This would allow suborbital vehicles (similar to Burt Rutan's *SpaceShipOne* discussed in Chapter 22) to be used for access to space, at a much lower energy cost than an orbital vehicle. The concept has been called a *Skyhook* and suborbital vehicles would fly up to a docking platform at the end of the cable, discharge passengers and cargo, then release and fly back. The electricity generated by the cable moving through the Earth's magnetic field could, perhaps, be used for electrodynamic propulsion to keep the elevator in position and to power the climber.[xxxix] Electricity could also be provided from solar arrays at the upper end of the cable. A technology demonstration test of electrodynamic propulsion was developed by a student team, but has not yet been tested in space.[127]

Space elevators are not limited to use on the Earth. They could also be used on the Moon, going from the surface to the L1 and L2 Lagrange points[xl] that are "lunar synchronous." They could be used to carry passengers and cargo and bring material from the Moon for construction in space. Elevators could also be used on Mars. Due to the lower gravity of the Moon and Mars, existing materials might have sufficient strength for those elevators.[128]

On Earth, highly conductive nanotube ribbons or wires would have other applications, as well. Presently, the electrical distribution system depends upon high-tension copper power lines to carry electricity from generating plants to end users. Of the electricity generated, almost 10 percent is lost as heat due to resistance in the copper lines, and this is trending upward as electricity is increasingly sent longer distances from generation site to consumer, in some cases thousands of kilometers (miles) apart. As of 2006, the estimated losses in the electrical distribution system totaled 1,630 billion kWhr worldwide. Upgrading to materials more conductive than copper, even if not super-

---

[xxxix] Electrodynamic propulsion uses forces resulting from electric currents in conductors as a space vehicle travels through a magnetic field. Irwin, Troy, "Orbital Applications of Electrodynamic Propulsion," Masters Thesis, Wright-Patterson Air Force Base, Ohio, School of Engineering, 1993, http://www.stormingmedia.us/31/3145/A314572.html

[xl] Lagrange points are the five locations where the gravitational attraction between two masses of dissimilar size (one is at least 25 times more massive than the other) "precisely cancels the centripetal acceleration required to rotate with them." As a result, an object such as a spacecraft would be in a stable orbit in a Lagrange point. "The Lagrange Points," Montana State University. http://www.physics.montana.edu/faculty/cornish/lagrange.html

conductive, would result in a significant improvement. Reducing the loss has the same effect as adding power plants, without the environmental or security concerns of additional power plants. In some locations copper lines *are* being replaced with superconducting lines of ceramic oxides of yttrium, barium, cesium, and other elements. To be superconductive, though, these lines must be cooled to 65 to 75K (-343 to -325°F) by liquid nitrogen (N_2). Development of a material that is superconducting at higher temperatures would be of great value. Carbon nanotubes might make that possible.[129]

Leaving Earth Orbit

THE FIRST CHALLENGE IS GETTING INTO ORBIT around the Earth, the next challenge is moving around in space. Rockets are one way, but there are other means of in-space propulsion, some of which do not require any propellant.

To keep the cable sufficiently taut, a space elevator requires a counterweight beyond the GEO position (or beyond the center of gravity for a LEO elevator), that would extend for thousands of kilometers (miles). This provides an interesting possibility for launching spacecraft. An object released from the climber as it ascends the cable would fall in an elliptical orbit about the Earth, until the climber reaches the GEO position, from where the object would remain in a circular orbit. If the climber continued beyond the GEO position, an object released from it would be traveling faster than needed to remain in a circular orbit and, therefore, would enter increasingly larger elliptical orbits until, beyond 47,000 km (29,200 mi), it would head away from the Earth, on a parabolic path to deep space. This provides a means of "launching" spacecraft without using propellant. The higher above the GEO position the greater the velocity. So, by careful selection of the release point for velocity and timing for direction, spacecraft could be sent on their way to anywhere in the solar system. This approach would work from a LEO elevator as well, even better in some ways. A GEO elevator must be aligned with the Earth's equator, but because the Earth is tilted relative to the plane of the ecliptic—the Earth's orbital plane, 23.44 degrees off of the Earth's equatorial plane—spacecraft released from the climber would be on a trajectory out of the ecliptic plane. A course correction might have to be made quickly if the destination was a planet or any other location in the plane of the solar system. This could be done by using the Moon's gravity with a swing-by maneuver or using on-board systems, but it is an added complication either way. (Alternatively, a spacecraft could be released so that its orbit around the sun crosses the orbit of the destination planet or other object, when it arrives at that crossing point.) A LEO elevator, in comparison, could orbit the Earth in the plane of the ecliptic. The portion of the LEO elevator above its center of gravity would serve the same purpose as the portion of the GEO elevator above the GEO position, with the added benefit that spacecraft could be released on more direct courses for their destinations.[130]

Another method of in-space propulsion that does use propellant (though not as much as chemical rockets) is the ion drive engine, a type of electric propulsion. This is a relatively new type of engine, but it was used on the *Deep Space 1* (*DS1*) mission launched in October 1998 to Comet Borrelly. *DS1* was a technology verification mission and was the first to use an ion engine as the primary means of propulsion (some communications satellites use ion engines for station-keeping). Launched in 2007, the *Dawn* spacecraft—on course to the asteroid belt between the orbits of Mars and Jupiter—is expected to arrive at Vesta (the brightest asteroid) in 2011 and Ceres (the largest asteroid) in 2015, after a flyby of Mars for a course adjustment. To accomplish this multi-destination mission *Dawn* also uses an ion engine, which will propel it to over 41,000 km/h (25,500 mph). Ion engines are powered by solar energy and use xenon as the propellant. They have relatively low thrust ("It takes four days to go from zero to 60 [mph]," according to Marc Rayman of the Jet Propulsion Laboratory) so there is less stress on the structure and payload, but they deliver 10 times as much thrust per kg (lb) of fuel as rockets. Ion drives can reach very high velocities by providing that thrust continuously, unlike chemical rockets that burn up their fuel in a few minutes and then coast. The *Voyager 1* spacecraft launched in

1977, the most distant manmade object that is presently heading out of the solar system, could be passed in about 10 years by an ion drive spacecraft launched today. Further advances in electric drives could improve their performance significantly.[131]

Another means of propellantless propulsion was the subject of Clarke's 1990 book, *Project Solar Sail*, with stories by several authors (including Clarke) about using the pressure of light from the sun to propel spaceships with enormous sails. The idea was proposed by Johannes Kepler in the 17th century, after observing that the tail of Halley's Comet (and other comets) always points away from the sun, even though the comet is not traveling directly toward the sun, indicating the presence of a "solar wind." While there is a solar wind of ionized particles, the pressure of light from the sun is greater (so it would be more accurate to call it a "light sail" rather than a solar sail) though still very slight. However, with a sufficiently large sail of a suitably strong and lightweight material (such as "Buckypaper"), even the low acceleration obtained from sunlight can result in very high velocities, even approaching the speed of light. A solar sail spacecraft launched in 2010 could overtake *Voyager 1* in about eight years. A solar sail could be used for interstellar travel, too, and it would even be possible to tack toward the sun, as sailing ships can tack against the wind. Some present spacecraft, including the MESSENGER[xli] spacecraft to Mercury, make use of the effects of sunlight pressure to aid in attitude control.[132]

On August 9, 2004, a Japanese experimental solar sail was launched on a suborbital mission to test two different deployment approaches. Both were successful. The Planetary Society attempted to test a solar sail in orbit in 2005, but their *Cosmos-1* spacecraft was lost when the Russian *Volna* rocket failed during launch. In 2008, NASA attempted to test a solar sail in space to be used as a drag chute to deorbit LEO satellites. The *Nanosail-D* spacecraft is about the size of a loaf of bread, weighing less than 4.54 kg (10 lb), but the sail opens to 9.3 m^2 (100 ft^2). That mission,

---

[xli] MESSENGER is the acronym for MErcury Surface, Space ENvironment, GEochemistry, and Ranging mission. http://messenger.jhuapl. edu

though, was lost due to failure of the SpaceX *Falcon-1* rocket. The Planetary Society is considering using spare *Nanosail-D* hardware for a new *Cosmos* mission.[133]

For missions that do not require rapid acceleration, solar sails or ion engines may be better ways to go. When faster acceleration is needed, however, the use of propellants may be preferred, especially if they can be acquired along the way, rather than having to be carried along for the entire trip. When going to the outer planets, for example, a variation on the typical liquid-fuelled rocket may be advantageous. Rather than using hydrogen or kerosene as the fuel, however, optimizing the engines to use methane would allow in-situ refueling. Methane is plentiful in the outer planets, including Jupiter, Saturn, Uranus, Neptune, and Pluto, and it could be produced on Mars using the almost pure CO_2 atmosphere. Only O_2 would need to be carried along (or H_2 for use on Mars), greatly reducing the overall mass of the spacecraft.[134]

Nuclear powered rockets can also provide the required acceleration. Nuclear energy has advantages, as well as drawbacks, with supporters and detractors depending upon which characteristics they focus on and how they weight the importance of the characteristics. Radiation concerns are one issue, but in space radiation from natural sources is a constant reality, from solar radiation and galactic cosmic rays. Because of this, the use of nuclear-powered spacecraft that stay in space may not be an issue.

The NERVA[xlii] project aimed to develop a nuclear-powered rocket during the same period that the Saturn rockets were being developed for the Apollo program. The NERVA would be significantly more efficient than the Saturn, as measured by the specific impulse or change in momentum per unit of fuel ($lb_f/(lb_m/s)$)—the thrust (force, lb_f) divided by the propellant consumption rate (mass/time, lb_m/s).[xliii] Specific impulse is measured in seconds (the other units cancel out), and the higher the

---

[xlii] NERVA stands for Nuclear Engine for Rocket Vehicle Application.

[xliii] Another way to look at specific impulse is that the I_{sp} is the velocity of the exhaust (distance/time, v_e) divided by the acceleration of gravity (velocity/time, g). Thus, in metric units, $(m/s)/(m/s^2)=s$.

number, the more efficient the use of propellant. The Saturn V rocket that sent Apollo to the Moon used kerosene (RP-1) and liquid oxygen (LOX), and had a specific impulse of 304 s. The Space Shuttle main engines, using liquid H_2 and LOX, have a specific impulse of 450 s. In comparison, rockets using H_2 as the propellant but fueled by nuclear energy have a specific impulse that is two to three times greater, 800 to 1,200 s. For a given size vehicle, therefore, a nuclear rocket can launch two to three times greater payload, or reach a destination more quickly. In the 1960s, some proposals for post-Apollo missions specified the NERVA II, that would be launched into orbit by a Saturn V, to be used for a mission to Mars. But those plans did not materialize, and the NERVA program was canceled in the 1970s. Although launching nuclear rockets from the Earth may have unacceptable risks, for vehicles that remain in space to be used for multiple missions nuclear rockets may provide distinct advantages.[135]

In comparison, ion engines have a specific impulse of 3,000 s or greater, though with very low thrust due to the small amount of mass ejected. Combining a nuclear power source with an ion engine would be advantageous for missions beyond the asteroid belt, where sunlight is considerably weaker. In 2003, NASA initiated *Project Prometheus* to develop a nuclear-powered ion engine for a mission to Jupiter, the "Icy Moons Orbiter" (JIMO), in 2017. The goal was to orbit three of Jupiters' moons in sequence for detailed observation. This mission would also be a demonstration of the propulsion concept and the first of more such missions to the outer planets. In 2005, however, the program was canceled.[136]

In 2004, a group of students at the California Institute of Technology, "excited by space exploration but uninspired by recent visions of NASA," developed a plan they call *Odyssey*. Their plan is a broad exploration of the solar system using a fleet of nuclear-powered spacecraft. They aim for a maiden voyage between 2025 and 2030, with a fleet of 10 spacecraft by 2050, for voyages to the inner planets, to Jupiter and Europa, and a tour of the asteroid belt. Their emphasis is on the means of reaching such places, though they realize that is not the only challenge. For such missions to be successful, the challenge of keeping the people alive and healthy must also be successfully met. In the harsh environment of space, how is that to be done?[137]

10

The Means, Part II: Living in Space

"Having people in low-Earth orbit is beneficial because we become better citizens for having been in space ... When you look back at the Earth from that vantage point, you can see that the atmosphere is like the shell of an egg or the skin of a potato—that's how thin it is; that's how little air there is to breathe. You also see how beautiful the Earth is—the colors, the waters, the continents. You learn to love our planet and you want to take care of it." — Eileen Collins, astronaut

LIVING IN A NEW ENVIRONMENT tests our abilities to adapt to different conditions. Whether going to a place with a different climate, moving from suburb to city or vice versa, or living in space, such changes bring new challenges. The environment of space presents hazards we are not yet adept at handling, but with every human space mission we gain experience.

• • •

B Y THE END OF 2001, IT WAS APPARENT that something was amiss with the thermal control system on board the *ISS*. *Destiny*, the U.S. laboratory module, launched on February 7, 2001, had been in operation for less than a year, but the pH of the circulating coolant had dropped below the specified range and there were traces of ammonia in the coolant. The lower pH allowed microorganisms to grow that could clog the filters and could induce corrosion of metal parts. The presence of ammonia could indicate a leak in the heat exchanger with the external ammonia coolant loop to the space radiators. Neither situation was good and a leak could threaten the mission. If the condition had continued to worsen, the U.S. portion of the *ISS* might have had to be abandoned.

Almost four years earlier, on January 6, 1998, I was assigned to be Lead Engineer for constructing a full-scale simulator of *Destiny's* internal thermal control system (ITCS). The simulator was to duplicate the heat transfer and coolant flow characteristics of the flight ITCS. A sub-scale simulator was also constructed to provide information on fluid chemistry changes, material corrosion, and microbial activity over extended periods of time. The subscale simulator was set up to operate continuously for three years and was referred to as the "Fleetleader" because it was expected that any problems would show up here before they would appear in the flight system. There were three general goals for these simulators: identify potential failures of the ITCS before they occurred on the *ISS*, support planning to avoid problems (such as scheduling the operation of the various experiment payloads to prevent overloading the ITCS), and troubleshoot problems that did arise. Both simulators had to be operational before *Destiny* was launched. Considering all that needed to be done this was a fairly tight schedule.[138]

Heat is a concern for the *ISS* and for all spacecraft. In the vacuum of space heat is transferred only by thermal radiation, since there is no fluid for convection to occur or direct physical contact for conduction to occur. In space, the sun's radiation is intense and when there is no atmosphere to moderate the temperature, the sun-facing side is very hot, while the side away from the sun is extremely cold. On the Moon, for example, the surface temperature ranges from 123°C (253°F) to -233°C (-387°F). For the *ISS*, and other spacecraft, the thermal control system is designed to provide the moderating effect of a planetary atmosphere. On the exterior of a spacecraft, multi-layer insulation (MLI) reflects most of the thermal radiation and minimizes the amount of heat that can be conducted to the spacecraft through the insulation. The MLI also reduces the heat loss on the shadow side to maintain a more even temperature within the spacecraft. For habitats in space—because of equipment, lights, and the people inside—excess heat is generated which must be removed continuously. This is a challenge that, out of necessity, was solved early in the space program. When you look at a picture of the *ISS*, you will notice the large solar arrays immediately. Less obvious are the radiators, smaller and with an accordion-fold pattern, usually oriented perpendicular to the solar arrays and sometimes in their shadow. The Space Shuttle, too, has radiators, and one of the first tasks performed when a Shuttle reaches orbit is to open the payload bay doors to deploy the radiators.[139]

For the *ISS*, the heat is carried to the radiators by circulating ammonia, which picks up heat from a heat exchanger mounted on the outside of *Destiny*. That heat exchanger transfers heat from the aqueous solution circulating through the ITCS inside the modules to the ammonia in the outside loop while maintaining physical separation of the fluids. The concept is rather simple, but complexities arise due to the operating conditions, and for that reason a ground simulator was needed.

The Fleetleader had been completed and started on September 5, 2000, and samples of the coolant were collected each month for analysis. In early 2001, validation of the full-scale simulator was in progress. As *Destiny* was prepared for launch, tests of the systems, including the ITCS, were performed at the Kennedy Space Center to ensure proper operation. The simulator at Marshall was then tested in a similar manner and the results were compared. As necessary, we adjusted the operation of the simulator until the results matched *Destiny*. We were ready to support flight operations.

In September 2001, the Fleetleader was steady as a rock—the conditions were virtually identical with the starting conditions. Meanwhile onboard the *ISS*, the pH was dropping, microbial growth was increasing, and evidence of corrosion was showing up in the samples returned for analysis. For some unknown reason, the Fleetleader was not leading. Despite our meticulous planning and construction of the facility, we had missed something critical or else something had failed in *Destiny*. Either way, we needed to determine what was happening and correct the problem before the situation worsened.

By late September 2001, the microbial growth was worrisome enough that quick action was called for. The simplest response was to add a biocide—silver ions. The question was how to do that, since provisions to add biocide had not been included in the ITCS design. After much discussion, we decided to place silver phosphate powder in an ITCS filter and have the astronauts perform a filter replacement procedure, something that they had practiced as part of their *ISS* mission training. But we needed to be sure that no unexpected problems would occur. The ITCS simulator would have its first chance to prove its worth. The necessary arrangements were quickly made and the procedure was performed before the end of October. The results were better than expected and on the next Space Shuttle flight, in November, two filters with biocide, one for each coolant loop, were taken to the *ISS*. The astronauts installed the filters, and coolant samples collected afterward showed that the microbial population was reduced to the desired level. A potentially dire situation was resolved by a relatively simple maintenance procedure, but we still did not know *why* the pH was dropping or why ammonia was showing up in the coolant. We had more work to do.

The ammonia concentration in later samples was not increasing from the earlier levels. This was good news, indicating that the source was not a leak with the ammonia loop, which would have caused the level to steadily increase, but it was not clear what other source there could be. Samples of the *ISS* atmosphere also showed the presence of low levels of ammonia, and for a time there was a concern that the ammonia was going from the coolant into the atmosphere. But ammonia is also a metabolic byproduct of people—we offgas. The trace contaminant control subsystem continually removes that ammonia from the atmosphere and, due to its solubility in water, ammonia is also removed by the humidity control subsystem. Since the ITCS coolant loop is pressurized, it was not expected that significant amounts of ammonia could get *into* the coolant *from* the atmosphere, especially with such low concentrations in the atmosphere and with the assumed low permeation rate through the Teflon hoses. However, the final conclusion was just that—even though the ammonia concentration in the atmosphere is low and the ITCS coolant lines are pressurized, the ammonia follows *its* pressure gradient, not the total pressure gradient, and permeates the Teflon hoses into the coolant. This was simply the background concentration. Ammonia was no longer a concern.

The low pH was still a concern, since it would increase the rate of corrosion of metal parts. By early 2002, the pH had leveled off as the coolant found a new equilibrium condition. The situation was no longer an emergency, but it was still a mystery. We considered everything from contamination of coolant samples to manufacturing issues to microorganisms in the coolant to permeation of CO_2 from the atmosphere through the Teflon hoses. Calculations performed during the ITCS design had indicated that CO_2 permeation through the hoses would not be significant, and experience with *Skylab* in 1974 indicated no concerns with permeation through the hoses.[xliv] The CO_2 concentration in the *ISS* atmosphere, however, is as high as 0.4 percent, over ten times higher than normal on the Earth. To rule out the possibility of CO_2 permeation, the Fleetleader was modified by enclosing the Teflon hose[xlv] and injecting pure CO_2 into the enclosure. Within days the pH began noticeably decreasing and within three months had dropped to the level in the *ISS* ITCS. We removed the CO_2, and the pH began increasing. We injected a mixed gas duplicating the composition of the *ISS* atmosphere, and the pH held at the *ISS* ITCS level. The CO_2, like the ammonia, was permeating from the atmosphere into the aqueous coolant. The mystery was solved. The higher rate of corrosion would shorten the life of metal components, but it seemed unlikely that the pH would drop further and the present corrosion rate was determined to be acceptable.

These issues with the ITCS show the unexpected ways in which living in space is different from living on Earth. We learned some important lessons—for one, a higher atmospheric CO_2

[xliv] The longest *Skylab* mission lasted 84 days and the total time it was occupied was less than six months. If *Skylab* had been operated for a longer period, the permeation issue may have been revealed, although the pressure and composition of *Skylab's* atmosphere were different from the *ISS*. For *Skylab*, the total pressure was 34.5 kPa (5 psia) with 72 percent O_2 and 28 percent N_2.

[xlv] One large Teflon hose (2.54 cm (1 in) diameter and 3.66 m (12 ft) long) had been used instead of numerous smaller ones, but the surface area was representative.

concentration can, over time, have significant adverse effects. We also learned, yet again, that it is important to test under relevant conditions. Experience with ground testing and *Skylab* was not sufficient to predict how the *ISS* ITCS would respond under operational conditions. Avoiding these problems requires thinking "outside of the box," and realizing that in a new situation, past experience may be inadequate to predict results. These were cheap lessons, unlike the Apollo fire and the *Challenger* and *Columbia* accidents. What you don't know can, indeed, hurt you.

Learning to live in space challenges us to look at problems in new ways, to gain greater understanding of how things work. When it comes to space flight, as with other areas of life, details can matter greatly. This is why "viewgraph engineering" is a dangerous practice, since the details often cannot be included on a presentation chart. Decisions may then be made with insufficient information. Knowledge and understanding, and recognition of where either are lacking, are essential to ensure that good decisions are made.

THE *ISS* IS A VALUABLE STEP toward becoming a space-faring civilization. Much has been learned not only from the experiments that have been performed, but also from the systems on board, and not only from the things that have gone well, but, perhaps more so, from problems that have become apparent during operation. Living in space is a very new activity, and there are aspects that do not become clear until we spend time there. The total time of all astronauts/cosmonauts/taikonauts in space is, perhaps, 150 person-years, about two lifetimes.[xlvi] With the completion of the *ISS* assembly and increase in crew size to six, the total time in space is increasing more rapidly, but our experience base is still quite low.

[xlvi] Accurate information on the total is hard to come by, but in 1999 the total U.S. astronaut time in space was about 150,000 hours (17 years). The USSR operated the *Salyut* and *Mir* space stations since the 1970s, acquiring considerably more time. Since the *ISS* became operational in 2000, crews of 2 or 3 have continuously lived on board.

In addition to thermal control, there are other challenges to living in space. As identified by Mike Griffin, former Administrator of NASA, during testimony to the U.S. Congress, "The high-priority areas are going to be space radiation, health and shielding, advanced environmental control and monitoring, advanced EVA activities and support of those, human health and countermeasures, life support systems, medical care for exploration and human factors, medical research with human subjects and microgravity validation of the environmental control and life support systems." NASA's *Constellation* program, which aimed to return astronauts to the Moon by 2020, provided a focus for advancing all of these areas. The cancellation of the program in 2010 will require readjustments, but the challenges will still need to be addressed to support the utilization of space. Of the high priority areas, radiation is, perhaps, the most challenging.[140]

Near the Earth, the magnetic field and the Earth itself shield astronauts from much of the radiation in space. Going to the Moon, and beyond, is a different matter. Of greatest concern are the relatively constant galactic cosmic rays (GCRs) and the transient solar particle events (SPEs). GCRs are very high energy particles that can come from any direction, but they occur at a low frequency, so the danger they pose is one of cumulative radiation exposure over the duration of a long mission. In comparison, an SPE is due to the most violent of solar flares and produces X-rays (which reach Earth in minutes), energetic particles (hours), and solar plasma (days) and can kill an unprotected person in a single burst. Monitoring of solar activity can provide a couple of hours advance warning, allowing astronauts time to take precautions. Those precautions could include taking refuge in a sheltered location with radiation-resistant walls in the spacecraft and orienting the spacecraft to provide the greatest protection by the vehicle structure or fuel or water tanks. Water is very good as a shield, because the water molecules don't fragment the way that larger molecules do. Hydrogen also is very good as a shield for the same reason and plastic polymers that contain large amounts of hydrogen also might be a useful material for shielding. In addition to shielding by mass, another possibility is an active electromagnetic field that would perform the same function as the Earth's magnetic field. The challenge there is providing a sufficiently strong field to block the most harmful radiation while ensuring that there are no adverse effects on the people or equipment.[141]

Better understanding of the sun and the radiation hazard may also lead to other mitigation methods. Two Solar-TErrestrial RElations Orbiter (STEREO) spacecraft were launched on October 26, 2006 to monitor the sun for the occurrence and characteristics of "coronal mass ejections" and other events. Orbiting the sun— one ahead of and one after the Earth—the twin spacecraft enable detailed long-term three-dimensional observation of the sun.[142]

For human missions to Mars, radiation exposure is a major concern. The radiation environment for such a trip was monitored by NASA's *Mars Odyssey* spacecraft that arrived at Mars in early October 2001. It found that along the way the radiation may not be as severe as previously thought, with the radiation exposure for a Mars crew during a 3-year mission well within the recommended career limits set by NASA. A network of radiation sensors in the inner solar system is needed to warn of impending solar storms so that space travelers can take precautions to minimize exposure. To further minimize the dangers of radiation, research at the NASA Space Radiation Laboratory in Upton, New York, is aimed at improving radiation countermeasures. These efforts will increase understanding of the radiation environment and enable us to develop better countermeasures.[143]

Moon dust has romantic connotations, but when living on the Moon it will be a major concern. The Moon is covered in a layer of regolith—a "stone blanket"—consisting of particles ranging in size from fine dust to boulders, with most about sand grain size. In the absence of the weathering processes common on Earth, the regolith particles are typically angular and abrasive, and electrostatic charges make them "sticky." For the Apollo missions, the dust was not a severe problem, although after three days of surface operations, dust was starting to affect the operation of the space suit joints and was showing up in the Lunar Module. For longer stays, an effective means of dust control will be essential. One way to control dust is to fuse the particles by high-powered microwaves to form solid blocks or make roadways. Fused regolith could also provide a stable surface for construction, reducing the infiltration of dust into the habitats, and could be used to form the radiation protection for the habitats. Moon dust may also be a source of oxygen,[xlvii] suitable for breathing and for fueling rockets, and, when nuclear fusion becomes viable for power generation, the helium-3 present in the regolith could be mined and brought to Earth.[144]

Other challenges to settling space include the physiological effects of zero or reduced gravity such as the loss of bone mass, temporary decline in immunity in microgravity, reactivation of dormant viruses (e.g., herpes and chicken pox), and increased virulence of some pathogens in space; psychosocial pressures, especially for smaller groups on extended missions; and the distance from well-equipped medical facilities. Improved space suits for use in space, on the Moon, on Mars, or on other planets or moons will also be essential. Improvements in the reliability and maintainability of equipment, especially computers, will be needed, and, of course, reliable and low-cost life support systems are critical for space habitats. NASA technology development programs are addressing each of these challenges.[145]

JUST AS DA VINCI'S IDEAS FOR AIRCRAFT could not be realized until the necessary technologies were available, so, too, utilization of the energy and resources in space and settlement of space will not occur until the enabling technologies have been developed. As may be evident from the examples in this book, major advances frequently build incrementally, and less often are the result of radical breakthroughs. To achieve such advances, the unceasing

[xlvii] The NASA Moon Regolith Oxygen (MoonROx) Centennial Challenge program offered a prize of $250,000 to the first team to extract a quantity of breathable oxygen from simulated lunar soil. The prize was first offered in 2005 with an expiration on June 1, 2008. The prize was not claimed.
"NASA Announces New Centennial Challenge," NASA Press Release 05-128, May 19, 2005, http://www.nasa.gov/home/hqnews/2005/may/HQ_05128_Centennial_Challenge.html

pursuit of technical excellence and "thinking outside the box" are essential. Though the technical challenges are considerable, so, too, are our capabilities. With appropriate efforts, the best ideas will be available to all who can best use them. We have the technology or soon will, but technology is not sufficient. We must also work together and coordinate our activities, which raises another set of challenges.

11

The Means, Part III: Co-opetition

"Great discoveries and improvements invariably involve the cooperation of many minds." — Alexander Graham Bell

FOR SUCCESSFUL COMPLETION OF MAJOR ENDEAVORS it is essential that the groups involved cooperate, even when they otherwise may be competitors. Such cooperation among business competitors has been called "co-opetition." As in the realm of business, efforts in space have benefited from cooperation among competitors— between countries as well as between companies—and future space efforts will also benefit from co-opetition.

• • •

In the early 1990s, the Space Station *Freedom* program was in deep trouble. Following the *Challenger* accident, costs increased despite repeated efforts to reduce them by reducing capabilities. A joke at the time was that *Freedom* would be renamed Space Station "Fred" to reflect the descoping. The SEI program also lost support by the early 1990s when the cost estimates became too great. NASA was facing the real and potential loss of its biggest programs.[146]

In June 1993, the funding to continue *Freedom* was approved by a margin of one vote in the House of Representatives. The program had survived, barely, but in late 1993 it was reorganized to bring in Russia as a partner. *Freedom* was redesigned to include Russian modules as key components and was renamed the *International Space Station* (*ISS*). Although this added complexity, the reorganization also provided additional rationale for the program.

The Russian *Mir* space station was operational at the time, the latest of several space stations that the USSR had launched starting with *Salyut 1* in 1971. These space stations gave the Russians considerable experience with long-duration stays in space, including for more than a year. (The longest single stay was 438 days by Valery Polyakov in 1994 and 1995.) In comparison, the longest stay by Americans was 84 days, by the third *Skylab* crew, twenty years earlier. The new agreement included NASA missions to *Mir*, which allowed the U.S. to gain long-duration experience in space years earlier than otherwise possible and to acquire direct information on the physiological effects and other aspects of living in space for extended periods. From 1995 to 1998, seven U.S. astronauts lived aboard *Mir* for periods from 128 days to 188 days, with Shannon Lucid being the American record holder.[xlviii] This experience was invaluable to NASA.[147]

Joining the *ISS* also gave the Russian space program a boost. Construction of two modules for *Mir 2* had stopped when the USSR fell. These were completed to serve as the core of the *ISS*. Since Russian and U.S. modules would be joined, it was essential that each side have sufficient understanding of the others' systems to ensure compatibility. Toward this end, in October 1994, translated documents describing the Russian life support devices were received at Marshall and distributed to the appropriate subsystem engineers.

As we learned more about the *Mir* life support system, it was apparent that there were some important differences in the way that the Russian engineers designed their systems from the way that we designed ours. For example, the biocide used to keep drinking water potable in the U.S. modules is iodine,[xlix] whereas the Russians use silver. Each works well, but if iodinated water is mixed with water containing silver ions, the result is the formation of silver iodide, a solid that has no biocidal properties. This was important to know so we could ensure that the waters remained

---

[xlviii] Russian cosmonaut Sergei Krikalev holds the world record for the most time in space, with a total of over 803 days during six space flights.
[xlix] For U.S. municipal water systems chlorine is used, but in the closed volume of a space habitat even small amounts of chlorine released into the atmosphere can cause problems. So, iodine, that is less volatile, was selected.

completely separate. A more fundamental difference between the U.S. approach and the Russian approach is that while each includes backup equipment for critical functions, the U.S. approach usually is to have a duplicate of the primary device, whereas the Russian approach usually is to have a backup using a different technology. For O_2 generation, for example, both the Russians and the U.S. use water electrolysis devices.[1] As a backup, however, the Russians use a simpler, though non-regenerable, technique, referred to as a "candle" because it uses compounds that produce an exothermic reaction while releasing O_2. Unlike water electrolysis, the candles use no power and can be stored for extended periods yet still be ready to use at a moments notice.

To ensure compatibility of both life support systems, we needed information about the Russian system and they needed information about the American system. However, there was no Russian manual on "how to design a life support system" that was comparable to *Designing for Human Presence in Space* that describes how the life support systems of U.S. spacecraft were designed (see chapter 21). Though the descriptions of specific devices were helpful,[li] a comprehensive description was needed of both systems. By the end of 1994, I was assigned to prepare a report for that purpose.[lii] The resulting document helped to ensure that the life support systems would work together properly and, as the *ISS* construction has progressed, no major problems have arisen

---

[1] In addition to water electrolysis, O_2 is also available from storage tanks on the U.S. side. This is O_2 "supply" available in an emergency, but is not "generation" of O_2.

[li] These documents were "machine translated" into English, and the translations were occasionally obtuse and required deciphering even more challenging than having the original Russian to translate. One of the more puzzling terms was "collectorless vacuum." After much pondering and considering the context of the usage, I determined that it probably meant "brushless motor." Common electric motors have copper brushes to conduct electricity to the rotor and sparks are generated as the brushes make contact. This is undesirable in space habitats, so motors that do not have these brushes are used.

[lii] Wieland, P.O., *Living Together in Space: The Design and Operation of the Life Support Systems on the International Space Station*, NASA, Marshall Space Flight Center, AL, NASA/TM-1998-206956, January 1998.

due to the differences in the American and Russian life support systems.

On an inter-governmental scale, the *ISS* is an example of co-opetition, and for all of the challenges encountered, it has increased peaceful international relations and has advanced the technologies necessary to live in space. The capabilities of the fully-assembled *ISS* will support future cooperative efforts, as well.[148]

COMPETITION IS FREQUENTLY THOUGHT OF AS THE DRIVER of innovation and in business "building a better mousetrap" is seen as the way to get ahead of one's competitors, but cooperation is also vital. In *Co-opetition*, Adam Brandenburger and Barry Nalebuff point out that "most businesses succeed only if others also succeed... your success doesn't require others to fail—there can be multiple winners." As an example, they refer to the relationship between Microsoft and Intel. "The demand for Intel chips increases when Microsoft creates more powerful software. Microsoft software becomes more valuable when Intel produces faster chips. It's mutual success rather than mutual destruction. It's win-win." Brandenburger and Nalebuff also say "[p]utting co-opetition into practice requires hardheaded thinking. It's not enough to be sensitized to the possibilities of cooperation and win-win strategies. You need a framework to think through the dollars-and-cents consequences of cooperation and of competition. ... Co-opetition recognizes that business relationships have more than one aspect. As a result, it can occasionally sound paradoxical. But this is part of what makes co-opetition such a powerful mindset. It's optimistic, without being naive. It encourages bold action, while helping you to escape the pitfalls. It encourages you to adopt a benevolent attitude towards other players, while at the same time keeping you tough-minded and logical. By showing the way to new opportunities, co-opetition stimulates creativity. By focusing on changing the game, it keeps business forward looking. By finding ways to make the pie bigger, it makes business both more profitable and more personally satisfying. By challenging the status quo, co-opetition says things can be done differently—and better."[149]

Co-opetition makes use of game theory[liii] to move beyond overly simple ideas of competition and cooperation to reach a vision more suited to current opportunities. Game theory was applied to warfare during World War II to locate Nazi submarines, and is now commonly applied to business. While game theory may conjure up images of winners and losers—the zero-sum game—it applies just as well to positive-sum—or win-win—games.

It is also possible to apply game theory to governmental interactions during times of peace and to space activities. When one country has something that another country wants—a resource, a technology, a capability—there are a number of ways by which the desired thing can be obtained by the other country. While war, espionage, or theft have all been used in such situations, so, too, have trade, joint activities, and treaties—examples of co-opetition—by which all parties expect to benefit.[150]

The space age began as a high stakes competition between the U.S. and the USSR. But even while the competition was grabbing the news headlines, cooperative space activities were occurring. On December 20, 1961, the United Nations General Assembly adopted Resolution 1721, stating that global satellite communications were to be available on a non-discriminatory basis. On August 31, 1962, President Kennedy signed the Communications Satellite Act, with the goal of establishing a satellite system in cooperation with other nations, and on August 20, 1964 the International Telecommunications Satellite Consortium (INTELSAT) was established with 11 participating countries. INTELSAT's first communications satellite, *Intelsat 1* (aka *Early Bird*), was the world's first commercial communications satellite placed in GEO, on April 6, 1965. Later additions resulted in a "constellation" of communications satellites, fulfilling Clarke's vision of continuous global coverage. By 2001, INTELSAT had over 100 member countries and operated dozens of satellites. As a mature telecommunications system, it was

---

[liii] Game theory is a branch of mathematics that describes behavior in strategic situations, where the success of one individual's choices depends on the choices of others.

decided that INTELSAT could be operated as a business. On July 18, 2001, after 37 years as an intergovernmental organization, INTELSAT was sold and became a private company (Intelsat, with lowercase letters), with the proceeds distributed among the member countries according to their respective use of services. As of 2007, Intelsat operates 51 satellites.[151]

Other space ventures involving international cooperation include Russia's spacecraft to land on Phobos, one of Mars' two moons, after a launch scheduled for 2011. This will be Russia's first mission to Mars since the *Mars-96* spacecraft, which was lost due to a launch failure. The name of this spacecraft is *Phobos-Grunt* (pronounced "groont," Russian for "soil") and, among other tasks, it will scoop up a sample of Phobos' soil for return to Earth. Other countries are also participating in the mission, including France, China, and the U.S. (an experiment from the Planetary Society is included). To the Russian engineers and scientists working on this mission, the international participation is reminiscent of the highly successful *Vega 1* and *2* missions to Venus in 1984. After that mission, according to Vechaslav Linkin, a scientist at the Russian Space Research Institute, "there was an impression that we can achieve so much through cooperation." Anatoly Perminov, a Russian space official, suggests international cooperation to address the threat of asteroid impacts, specifically *Apophis* in 2036. Such cooperative efforts benefit everyone.[152]

CO-OPETITION IN COMMERCIAL SPACE ACTIVITIES is also important. With the success of Burt Rutan's *SpaceShipOne* in winning the Ansari X Prize in 2004, interest in suborbital tourist flights is increasing. Richard Branson has ordered a fleet of *SpaceShipTwos* from Rutan to begin offering public flights in 2011 through his new company, Virgin Galactic. Spaceports to support such flights are being developed in several states in the U.S. and are being considered in Singapore, Australia, and the United Arab Emirates. The operators of these various spaceports are competitors, but they have also formed an organization, the Personal Spaceflight Federation, to promote their common interests. According to Alex Tai, chief operating officer for Virgin Galactic, and the federation's chairman, "it was essential for even rivals to cooperate in the formulation of policies for dealing with accidents and government regulations." He noted that government agencies such as the Federal Aviation Administration, which has jurisdiction over such space flights in the U.S., "prefer to deal with a unified industry group on the larger issues surrounding private-sector space travel." "Spaceport managers themselves [are] looking forward to a second stage of suborbital travel, in which travelers can take rocket-powered flights from point A to point B. For that reason, spaceport authorities are interested in interoperability as well—a complex blend of cooperation and competition familiar to dot-com veterans under the term 'coopetition.'"[153]

THE SPACE AGENCIES OF COUNTRIES around the world, to a considerable extent, practice the principles of co-opetition. Numerous joint programs are in progress (including Earth observation and climate monitoring), capabilities and facilities are shared, and plans are underway for future cooperative activities. This trend is growing. Looking ahead to a time when multiple countries have habitats in space, NASA hosted a meeting in 2006 with representation from 14 space agencies (including from Russia, China, Japan, India, Ukraine, Korea, Australia, Canada, and European countries) to discuss international interests and to coordinate activities where feasible and beneficial. The resulting report, *The Global Exploration Strategy: The Framework for Coordination*, released in May 2007, expressed "a shared vision of the role of governments around the world to extend human and robotic presence throughout the solar system." Following that initial meeting, a formal organization was formed, the International Space Exploration Coordination Group (ISECG), which met in Berlin in November 2007. The Framework document reads, "[t]he ISECG is open to space agencies which have or are developing space exploration capabilities for peaceful purposes and which have a vested interest to participate in the strategic coordination process for space exploration." "In sum, it is not an exclusive club of the 14 agencies that developed the Framework document."[154]

One example of co-opetition in space is the agreement, in the 1990s, for joint activities between the U.S. and Russia. Before NASA could undertake Space Shuttle missions to the *Mir* space station, a means of docking had to be developed. *Mir*, of course, had docking stations for the *Soyuz* and *Progress* spacecraft. The Space Shuttle had a connecting passage for the *Spacelab* modules carried in the cargo bay. But when the respective spacecraft were designed, no one considered that the Space Shuttle would ever dock with *Mir*. A special docking adapter had to be created to allow that. In the future, if such mismatched hardware were used, there might not be time to develop adapters to allow docking. The ISEGC realized that could be a serious problem for other types of interfaces as well and is taking the initial steps "to identify critical space-infrastructure interfaces—such as connections among spacecraft, lunar rovers and lunar habitats—that, if standardized, would increase opportunities for international cooperation." Standardization would increase safety as well. Standards establish uniform engineering or technical criteria, methods, processes and practices, and make it possible to interchangeably use electronics, for example. For habitats on the Moon or Mars or in space, the use of standardized docking systems, common atmospheric standards, and communication protocols will improve safety and increase the opportunities for cooperative activity. The ISECG will identify the critical interfaces that should be standardized. In one possible scenario, a Japanese transporter may need to connect with a European habitation module and receive electricity from an American power station.

Space exploration activities are also being coordinated between space agencies. This will minimize redundancy and help to increase the data return and infrastructure development. Government policies, regulations, legislation, and international agreements are all critical aspects. Activities in space, as well as actions addressing energy and environmental issues, will require the coordination that is integral to co-opetition. What could we achieve with even greater coordinated action?[155]

12

Convergence

"Coming together is a beginning. Keeping together is progress. Working together is success." — Henry Ford

AN ACHIEVEMENT WILL OCCUR when the vision, motivation, and means converge to accomplish it. If any one of these aspects is missing, or too immature, then the achievement will not occur. If, for example, the motivation and means are present, but a vision of the end result is not yet sufficiently developed, then action is stunted or ineffective. When a vision is present but seems to be too fantastic to ever be realized, it often is because the means have not matured sufficiently. Our current situation regarding utilization and development of space is a mixture of these situations. There are numerous variations of the general vision for space activities, some of which address issues on Earth, but no single vision has acquired broad appeal and, as the previous chapters indicated, we are still developing the means to carry out any of them in an effective manner. However, the time of convergence—when the vision is clear, the motivation is broadly felt, and the means have matured— is nearing.

• • •

IN 2007, MICHAEL GRIFFIN, while Administrator of NASA, presented his view of the next 50 years of space activities. Projecting that NASA's budget will remain constant, at about $17.3 billion/year (FY2008), his prediction was for the Vision for Space Exploration to return astronauts to the Moon by 2020, beginning development of a permanent outpost, and in the 2020s, to "begin the Mars effort in earnest" with the first landing in 2037. His forecast "does not require a course change from present understandings, nor does it require extensive development of costly new technology. It is a logical, incremental, stable, sustainable plan that can be executed with realistically attainable budgets." Griffin's forecast is intentionally conservative, assumes gradual changes, and is based on stability not only of budgets, but also of goals and methods. As such, it may be the most reasonable forecast possible. However, it does not clearly address issues on Earth and it is a linear approach. One lesson of history is that, often, linear change is eventually overwhelmed by innovation and radical change.[156]

THERE IS A LAG FOLLOWING A SEMINAL EVENT until a vision represented by that event can be achieved. In the case of a breakthrough development, for example, the possibilities represented by that breakthrough take time to comprehend. Or the lag may be just the time it takes for incremental advances to accumulate sufficiently. But the lag is a critical period, during which a vision is clarified, the motivation gains support, and the means are developed to accomplish the vision. When a vision addresses fundamental societal issues, there is the potential for a major achievement, and when the vision, motivation, and means mature and converge, such achievements can be realized with astounding rapidity.

For both the settlement of the American West and the rise of aviation, after the key seminal event there was a period of several decades of essentially linear, incremental change. Forty years after the Lewis & Clark expedition, settlement had only gradually extended into the Louisiana Territory, along the Mississippi and Missouri Rivers, so Jefferson's expectation that it would take a thousand generations to settle the West still seemed likely. Forty years after the Wright brothers' first flight, while aviation was quite familiar, airline travel was still for the well-to-do and relied on slow, noisy piston-engine aircraft. The developments that would "change the game" had not yet matured.

The vision, motivation, and means were developing, but convergence would not occur for another decade or two. Settlement of the West did not "ramp up" until 60 years after Lewis & Clark. Similarly for aviation, it was over 50 years after the Wright brothers that jet engines—and the airport and air traffic control infrastructure—allowed air travel to "take off." Then the changes were exponential regarding migration westward and the number of airline passengers, respectively, and within another 30 years the visions were fully realized.

For space enthusiasts, the grand vision for space activities is settlement of the solar system, utilizing the resources found there and ultimately venturing further into the universe. As generally described, this is not a compelling vision for most people. For such space activities to become a widely held vision to be pursued now, and not just an interesting idea for some distant time, it is necessary that they be considered an integral part of addressing societal issues. For that reason, the vision needs to focus on those issues and how it serves them.

Especially when funding is difficult to come by, with pressures to reduce funding for activities considered to be less important, it is necessary to ensure that the funding that is received is spent well and in ways that effectively address issues. Several proposals for U.S. space policy have been presented recently that recommend shifting NASA's priorities to address societal issues and to promote the commercial space industry. In 2001, Tom Rogers, a physicist and space industry supporter, suggested that LEO activities be the domain of commercial space companies —which are eager to have that role—while NASA focus further out, to activities on the Moon and Mars. In 2009, in a report co-authored by George Abbey, former director of the Johnson Space Center, one recommendation is for NASA to focus on energy and environmental concerns. The common factor of these proposals is that a new vision is needed for the U.S. space program—one that is relevant to the issues of the 21st century.[157]

Addressing needs on Earth is the primary motivation for the vision presented in this book. If it seems unlikely that over the course of the 21st century the advances required to utilize space to address Earth-bound issues directly and to establish large settlements on the Moon or Mars or orbiting the Earth or sun could happen, recall from Chapter 3 the historical example of the great changes that occurred over one lifetime. The same rate of change can occur, and is occurring, regarding space activities.

JULY 20, 2009 WAS THE FORTIETH ANNIVERSARY of the *Apollo 11* landing on the Moon. We are currently at that pre-transcontinental railroad, pre-jet engine stage of space activities, so it may appear that major development is still only a very distant possibility. This will not be the case for long, especially if particular actions are performed, as described in the next part of this book. We are nearing the threshold of key developments and a period of exponential change. The "pause" in our advance into space— Gehman's "30 years without a guiding vision"—is almost over and the next decade is pivotal in preparing us to "cross the threshold."

The actions proposed in Part IV are relatively modest, but together they present a plan to utilize space to address issues on Earth while preparing for broader activity in space. The resulting technologies, infrastructure, and international agreements will provide the means for further activities, including establishing settlements in the solar system. By implementing the proposals in Part IV, by 2020 we will have not only a clear vision of how space activities can specifically address key issues, but also the means to carry out those activities.

Part IV

Creating the Future Now

"Earth is the cradle of humanity, but one cannot live in the cradle forever." — Konstantin Tsiolkovsky, 1911

• • •

SPACE ACTIVITIES COULD MORE DIRECTLY address energy, resource, and environmental issues, while enhancing international relations. In order to realize these possibilities, it is necessary to provide ready access to space, which will require a range of technological advances; to establish a compelling program of space activities, that inspires individuals and draws numerous countries together; and to promote commercial space endeavors, which will require international agreements and clear guidelines for activities in space. The government roles in these activities are important: performing activities that private enterprise cannot (or cannot yet) perform and supporting activities that private enterprise can perform, but that are not yet supported by a market. The fundamental research to acquire new understanding (that can be applied to develop new technologies) is often out of reach of private enterprise and, so, frequently must be sponsored, at least, or conducted by the government. The government must also take the lead regarding international agreements, especially regarding operations in space and legal considerations.

Seven proposals are presented in the following chapters, for actions that will enable us to acquire—by 2020—the knowledge, capabilities, and agreements needed to directly address critical issues through activities in space. The proposals are for actions that are readily doable (in some cases extensions of activities that are already being done) and do not require waiting for major technological breakthroughs to occur, while enabling such breakthroughs to be rapidly utilized when they do occur.

13

Leadership in Space

"Leadership is ultimately about creating a way for people to contribute to making something extraordinary happen." — Alan Keith, Genentech[liv]

"If your actions inspire others to dream more, learn more, do more and become more, you are a leader" — John Quincy Adams

"Now it is time to take longer strides—time for this nation to take a clearly leading role in space achievement, which in many ways may hold the key to our future on Earth. ... Space is open to us now; and our eagerness to share its meaning is not governed by the efforts of others. We go into space because whatever mankind must undertake, free men must fully share ..." — President John F. Kennedy, Special Joint Session of Congress, May 25, 1961

LEADERSHIP TYPICALLY INVOLVES either doing something first, doing something better, or doing something that others want to be a part of. The common aspect is doing something. Leadership is not about dictating to others but rather about having an inspiring vision and working toward that vision with the participation of others.

Leadership is essential to implement a vision. Someone must develop the goals, make the plans, and coordinate the actions to achieve the vision, preferably with the input and concurrence of everyone involved. This may especially apply to international efforts. Leadership also requires knowledge, so that must first be acquired.

• • •

[liv] Kouzes, James M., *The Leadership Challenge*, Jossey-Bass, 4th ed., 2008.

I N SYENE, EGYPT, THERE WAS A WELL in which the sun shone to the bottom on only one occasion each year, at noon on the summer solstice—the well was on the Tropic of Cancer, the line on maps indicating the furthest north that the sun is ever directly overhead. Eratosthenes, a Greek mathematician in the third century BCE, knew that his location (Alexandria, on the Mediterranean coast) was almost due north and he also knew the distance to Syene, as precisely as could be determined at that time. So, perhaps by measuring the length of the shadow cast by a vertical stick at noon on the summer solstice and then using his knowledge of geometry and knowing that the Earth is a spheroid, Eratosthenes calculated the circumference of the Earth—to within one percent of the actual value. This was a remarkable achievement considering his crude measurements, but he understood "the physics."[158]

Eratosthenes was a polymath with a questioning mind. He knew that the key to understanding was knowledge and he sought to acquire knowledge. Among his other accomplishments, he formalized the study of geography, even giving it that name, γεωγραφια, Greek for "writing about the Earth." Over the centuries since, scientific knowledge of the Earth has increased, replacing supposition and superstition, but, as is typical in science, the answers to questions lead to more questions.[159]

In 1952, the International Council of Scientific Unions proposed a coordinated effort to answer questions about the Earth and its environment. New capabilities, such as radar and rockets, enabled previously impossible explorations to answer previously unanswerable questions. The period from July 1957 through December 1958, when solar activity was at a maximum in the 11-year solar cycle, was designated International Geophysical Year (IGY), and 67 nations participated, including the U.S. and the USSR. Among the numerous firsts that occurred as part of the IGY, spacecraft were launched into orbit—*Sputnik 1*, the first spacecraft, and *Explorer 1*, which discovered the Van Allen radiation belt around the Earth. On June 30, 1957, as the IGY was about to begin, President Dwight "Ike" Eisenhower "expressed his belief that 'the most important result of the International Geophysical Year is that demonstration of the ability of peoples of all nations to work together harmoniously for the common good. I hope this can become common practice in other fields of human endeavor.'"[160]

In the same spirit, the year 1992, the 35th anniversary of the IGY and the 500th anniversary of Columbus' discovery of the Americas, was designated International Space Year (ISY) at the request of President Reagan in 1981. The theme for ISY, "Mission to Planet Earth," was presented by President George H. W. Bush on July 20, 1989, when he called for a major international effort to study the environment. Commemorating the 20th anniversary of the *Apollo 11* landing on the Moon, he reminded us of what the astronauts saw during their mission: "The Earth, blue and fragile, rising above the Sea of Tranquillity." The goal was "to better understand the complex interactions between land, water, air and ice and assess such threats as global warming, deforestation and ozone depletion."[161]

There have been other international years such as ones to study the sun and its effect on the space environment and solar system, including the Earth (International Heliophysical Year[lv] from 2007 to 2009), and the Earth's polar regions and the changes occurring there, especially regarding the accumulation of snow and melting of ice (International Polar Year,[lvi] also from 2007 to 2009).

The questions to be answered never end.

TO ANSWER QUESTIONS ABOUT THE EARTH AND MOON, our solar system, and the universe at large NASA has launched hundreds of

---

[lv] The goal of the International Heliophysical Year is to "determine the response of terrestrial and planetary magnetospheres and atmospheres to external drivers" (i.e., the sun), even considering the heliosphere as far as the interstellar medium (beyond the orbit of Pluto or Neptune). http://ihy2007.org/

[lvi] The International Polar Year is to study the changes occurring in the polar regions. The "three fastest warming regions on the planet in the last two decades have been Alaska, Siberia and parts of the Antarctic Peninsula. Thus the Polar Regions are highly sensitive to climate change." http://classic.ipy.org/ http://www.cbsnews.com/stories/2007/02/26/tech/main2516097.shtml

spacecraft since *Explorer 1* and, as of 2006, has 55 space science missions actively operating or capable of being operated. These include flyby and orbiting spacecraft, telescopes peering into the vastness of deep space; landers and rovers on Mars; and comet impactors. Many more missions are in the works to make more detailed observations and to bring back samples of the Moon, asteroids, and comets.[162]

Other countries (Russia, Europe, Japan, China, India, Brazil), too, have robust space programs, having collectively also launched hundreds of spacecraft, with missions to the Moon, Mars, other planets, asteroids, comets, and elsewhere. They are testing out new methods of propulsion and new instruments for gathering data and have plans for increased exploration of the solar system, including sending people to the Moon. China is the third country to develop the capability to launch people into space and, on September 27, 2008, during their third such mission, Zhai Zhigang became the first Chinese to walk in space.[163]

Leadership in space activities involves advancing knowledge and developing technologies in ways that others will want to participate. Promoting an IGY-type effort is one way to provide leadership. An international year-type program will obtain the greatest knowledge return for the effort, while minimizing the cost for each country participating. However, since many points of interest in the solar system may take more than one or two years to reach, a longer term effort is warranted to study the entire inner solar system out to the asteroid belt and beyond. Toward that end, an International Space Decade should be established with a goal of acquiring detailed knowledge of the solar system by 2020, so that better informed decisions can be made about the merit of particular space activities and how they could address needs on Earth.

Proposal 1: Implement an International Space Decade, from 2011 to 2020, to coordinate missions to obtain detailed information about locations throughout the inner solar system, especially as far as the asteroid belt. Such a coordinated, comprehensive program will also promote international cooperation.

14

Space Commerce

"[W]e would not have CNN, DirecTV™, XM Radio™, OnStar™, or Google Earth™ if it were not for ... space-based assets. ... Our daily weather forecasts would be far less reliable without earth observing satellites. Clearly, space is important to our daily lives, and it is in our economic interest to encourage further development of this 'final frontier' of business." — Edward Morris, U.S. Department of Commerce, Director of Space Commercialization[164]

WHILE THE GOVERNMENT HAS A LEADING ROLE in space activities, one goal should be to promote commercial endeavors in space. There are several ways in which this can be done, including making the results of government-sponsored research available for commercial applications, providing incentives for private enterprise to perform particular activities, and sponsoring or performing demonstration projects to reduce the technical and financial risk for companies to perform specific activities. These are already done to some extent, but current efforts can be expanded.

• • •

THERE IS A GROWING LIST OF SPACE MISSIONS that scientists, organizations, companies, and government agencies would like to undertake. The high cost of accomplishing these missions and the limited funding available mean that most of them will not occur anytime soon. NASA currently funds exploration missions primarily through the Discovery and New Frontiers programs, intended to contain costs by having funding caps—$425 million for each Discovery mission and $750 million for each New Frontiers mission. Missions costing more than $750 million are called "Flagship" missions and are exceptionally ambitious, and less frequent. Even with efforts to minimize the expense of current and planned space missions, the total cost for them is billions of dollars each year.[165]

One reason for the high cost of spacecraft is the way that they are designed and manufactured. Most spacecraft are individually designed and then assembled one unique spacecraft at a time, an expensive way to manufacture. The few exceptions to this situation reveal the potential to greatly reduce the cost of spacecraft. The experience with manufacturing the Shuttle SRBs, discussed in Chapter 4, shows that producing multiple copies of spacecraft can significantly reduce the time and labor cost of production for each one, and therefore the monetary cost, just as for other products. This desirable effect is even more apparent with some commercial ventures, such as Iridium and GlobalStar.

In the 1990s, the idea of using satellites to provide global telephone coverage from any place on Earth was pursued by several companies. One company, Iridium, had a plan to place a constellation of 77 spacecraft into polar orbits to provide full coverage of the Earth.[lvii] Launched in 1997 and 1998, these spacecraft are not simple, small satellites like the first *Sputnik* or *Explorer* satellites, but sophisticated vehicles weighing about 700 kg (1,500 lb), with advanced communications capabilities, designed to last for many years. By mass producing a single design and launching several spacecraft at a time on a single rocket, the cost per spacecraft was reduced substantially.[lviii] With an investment of $3.759 billion—that covered construction, launch,

---

[lvii] Since 77 is the atomic weight of the element iridium (which is also an element found in asteroids and serves as a marker of asteroid impacts) the idea, and the company, were called Iridium.

The number of satellites in operation was later reduced to 66, after determining that full Earth coverage could be achieved with that many. The total number built was 72, including spares and test satellites.

[lviii] Iridium spacecraft were launched five at a time on *Delta II* rockets, seven at a time on Russian *Proton* rockets, and two at a time on Chinese *Long March* rockets.

and operation for the first year—the average cost for 72 spacecraft was about $52 million.[166]

A similar venture, GlobalStar, uses 52 spacecraft in LEO, launched in 1998. Alenia Spazio in Italy manufactured the spacecraft, each weighing 450 kg (992 lb), at a production rate of up to one each week. A new generation of GlobalStar spacecraft are being launched beginning in 2009, with construction costs of $900 million (€ 661 million) for 48 spacecraft, an average of $18.75 million (€ 13.78 million) cach. Launch and operation add to the cost, but the total average cost for each spacecraft likely will be much less than the $52 million for Iridium's spacecraft. [167]

Further cost reductions are possible. In addition to mass production and launching multiple spacecraft on a single rocket, advances in miniaturization and new technologies can reduce the overall size and mass of a spacecraft and, therefore, the cost. Such "smallsats" hold the promise of providing communications, Earth imaging, and other capabilities at far lower cost than traditional spacecraft. The *Near Earth Object Surveillance Satellite* (*NEOSSat*) of the Canadian Space Agency is one of a new generation of microsatellites. With an expected launch date in 2011, *NEOSSat* is the size of a suitcase and weighs 65 kg (143 lb). From 700 km (435 mi) above the Earth, it will monitor the sky for NEOs and other spacecraft. At a cost of $12 million, *NEOSSat* points to a future where microsatellites become the standard. The use of more recent technology and off-the-shelf parts has sped up the miniaturization process.[168]

As part of NASA's New Millennium technology demonstration program, Goddard Space Flight Center developed the *Space Technology Five* (*ST5*) mission, with three micro-satellites launched to LEO on March 22, 2006. The new technologies included advances such as a communication transponder miniaturized to about the size of a chicken egg. As a result, each spacecraft weighs only 25 kg (55 lbs). The cost for this mission was $130 million for the three satellites, but the ultimate goal is to send dozens of micro-satellites into space for $1.5 million each. Other New Millennium missions include *Deep Space 1*, which successfully demonstrated the capability of xenon-ion propulsion and other advances. By testing these new technologies in space, the New Millennium program reduces the risk for other missions and enables those missions to be less costly or to accomplish more. Other advances in miniaturization will revolutionize GN&C[lix] with complete low cost inertial measurement units as small as 0.49 cm^3 (0.03 in^3) and will result in ion thrusters "no larger than a postage stamp that could power spacecraft the size of soup cans." The National Science Foundation is sponsoring development of "CubeSats" (that are not much larger than soup cans) to study space weather (the solar wind, etc.) and perform other specialized monitoring.[169]

One of the challenges to designing such micro-satellites is thermal control. The techniques used on larger spacecraft are bulky and heavy—one reason the spacecraft are so large—and are not suitable for small spacecraft. In 2008, however, Prasanna Chandrasekhar and his colleagues developed a thin film thermal material that controlled the temperature of a simulated spacecraft in laboratory testing. The material still needs to be tested in space, but the laboratory results are promising and successful application to spacecraft will increase the capabilities of smallsats.[170]

All of these developments together—mass production, multiple spacecraft per launch, miniaturization, and new technologies—can dramatically reduce the cost of spacecraft. Lower cost spacecraft would make many more missions affordable, promoting more mass production and lowering the costs even further. Designing spacecraft with standardized systems for communication, GN&C, power supply, propulsion, and other common functions would also support mass production, as would using standard suites of instruments (cameras, spectrometers, magnetometers, radiation monitors, solar flare detectors, etc.). By selecting from such standardized instruments, spacecraft can be customized for specific missions, much as cars are offered with a variety of engine and transmission configurations and other options. By designing the standard spacecraft to be able to carry an additional customer-provided payload or to be attached to a unique special-purpose

---

[lix] GN&C is the abbreviation for "guidance, navigation, and control."

spacecraft—similar to the way pick-up trucks can carry payloads or pull trailers—even one-of-a-kind scientific missions would benefit from mass production. Such efforts to reduce costs would also benefit spacecraft sent into deep-space in the solar system. These spacecraft would be performing the roles of the early explorers of a region, significantly increasing the return of data and our knowledge of the solar system.[171]

New types of missions would be possible, too. For example, multiple spacecraft outfitted with relatively small telescopes could be arrayed in formation and electronically linked to simulate a much larger telescope. Such a configuration could provide a "synthetic aperture" thousands, or even millions, of kilometers (miles) across, greatly increasing the resolution and possibly enabling direct viewing of Earth-type planets around other stars. Low-cost, mass produced spacecraft could also provide opportunities for profit, such as missions to search asteroids for rare materials. Or Google, Inc. may choose to add to its Google Earth, Google Moon, and Google Mars capabilities with Google Mercury, Google Ceres, Google Titan, Google Saturn, Google Pluto, and so on, using its own spacecraft. With low-cost landers and rovers Google could add surface-level images as well, enabling anyone to take a virtual tour of any of those places. (And to develop such spacecraft, Google is offering the $30 million Google Lunar X Prize to "the first privately funded team to send a robot to the moon, travel 500 meters, and transmit video, images, and data back to the Earth.")[172]

DRAMATICALLY INCREASING THE NUMBER OF MISSIONS could overwhelm our capability to monitor, process, and control them, however, so, improved methods of mission control are also needed to reduce the time to process data, the number of people required to monitor missions, and the need for specialized facilities. An infrastructure in space to enhance communications and navigation would support this goal. The *Mars Reconnaissance Orbiter (MRO)*, launched in August 2005 and operating in orbit around Mars, has been called "the first installment of an interplanetary internet," with a data transmission rate several times faster than a typical digital subscriber line (DSL), allowing vehicles on the surface to communicate through the *MRO* rather than directly with Earth. For operations on the Moon, a global positioning system (GPS)-type method has been proposed to aid in navigation. Extending these ideas to include interconnected communication and navigation systems throughout the inner solar system would aid communicating with, and monitoring of, spacecraft. With mass produced spacecraft, such a Solar System Positioning System or Interplanetary Internet would be affordable, and it would help to reduce the cost of other missions.[173]

To produce the most capable spacecraft at the lowest cost, the best ideas need to be combined. But the best ideas may have been developed by competing companies. This situation occurred in the early days of airplanes when manufacturers became embroiled in patent disputes that were not resolved until a cross-licensing process was developed. To avoid such a situation regarding spacecraft, a similar cross-licensing approach should be implemented now, utilizing the principles of co-opetition.

To promote development of mass produced, low cost spacecraft there needs to be a market for them. As has been done at other times, the government can be instrumental in stimulating the market by being a pioneer customer. One way for the government to do this would be to guarantee that a set number of spacecraft would be purchased during a particular time period. Such a guarantee would provide certainty for the companies that would build the spacecraft.[ix] For example, by guaranteeing that over the next decade 1,000 spacecraft would be purchased, at an average production rate of two spacecraft each week, more than one manufacturer could be supported, allowing multiple companies to provide spacecraft. The government would not need to agree to purchase 1,000 spacecraft, but, rather, would assure a minimum

---

[ix] A similar approach is taken by some countries, states, and cities to promote the use of renewable energy, by implementing long-term, guaranteed purchase contracts. Blake, Mariah, "The Rooftop Revolution," *Washington Monthly*, March/April 2009, http://www.washingtonmonthly.com/features/2009/0903.blake.html

level of production to reduce market risk. The government would be the "fallback" customer, ensuring that a minimum number of spacecraft are purchased each month. Initially the government would likely be the main customer, but those spacecraft could be used for a general survey of the solar system and for the initial interplanetary internet and solar-system-positioning-system, building the infrastructure to support further activities.

The capabilities of these spacecraft should be specified and include the capability to reach any location in the inner solar system (including the asteroid belt) within a reasonable period of time—perhaps within 12 months of launch—and to orbit the destination, if desired, not simply fly by. An additional design requirement should be that the spacecraft leave no debris in Earth orbit as they head for deep space and, if intended to remain in Earth orbit, be designed to be deorbited when they are no longer usable or needed. Sending spacecraft into deep space, away from Earth orbit, is more challenging and expensive than simply orbiting the Earth, but with goals of reducing the cost to less than $5 million/spacecraft for LEO operations, a cost goal of no more than $15 million each, including launch costs, for spacecraft that can reach any location in the inner solar system seems reasonable. At that cost, the number of potential customers would increase over time, and the government share would diminish as the in-space infrastructure is developed.

Similar guarantees could be made for spacecraft capable of landing on other planets or moons or asteroids—for perhaps 250 spacecraft over 10 years for no more than $40 million each—and rovers or sample return missions—for perhaps 100 spacecraft over 10 years for no more than $75 million each. The average annual cost over 10 years for all of these guarantees would be $3.25 billion. This is comparable to the current annual budget for exploration missions, but would provide many more missions, and the government would not be paying for all of them.

Proposal 2: Provide a government guarantee that over a decade at least 1,000 spacecraft—meeting specified performance and capability criteria—will be purchased (not necessarily by the government) for no more than $15 million each including launch costs. Comparable guarantees could be provided for landers and rover/sample return missions. This would support Proposal 1 while promoting commercialization and lowering the cost of space access.

15

Resources from Space

"If humans are ever to truly spread their wings in space, they must be nourished and sustained by space resources. ... Off-world resources can be transformed into oxygen, propellant, water, as well as used for construction purposes and to energize power stations." — Leonard David[174]

THE MOST RECENT SIGNIFICANT ASTEROID EVENT on the Earth occurred on June 30, 1908, when an asteroid hit the Earth's atmosphere over the remote Tunguska region of Siberia, Russia. Because of its density and speed, it plowed into the atmosphere, rapidly heating due to friction with the air, until it exploded with the force of a 15-megaton bomb. The explosion flattened 60 million trees, over an area 100 km (62 miles) across.[175]

Asteroids, obviously, have the potential to be extremely destructive. They are also a potential source of raw materials and some are rich in the platinum-group metals that are rare on Earth and very valuable. A much older and larger asteroid impact occurred at Sudbury, Ontario, Canada. This asteroid is now one of the richest nickel mines in the world. With suitable access to space, asteroids can be mined as they orbit the sun (or where they have hit the moon), to acquire the materials—including the rare elements, frequently required for batteries, fuel cells, and other energy-related technologies and high strength materials—for use on Earth.[176]

• • •

THE RESULT WOULD HAVE BEEN FAR MORE DEVASTATING had the Tunguska asteroid impact occurred over a heavily populated area. Out of concern that such an event may occur, the NASA Authorization Act of 2005 specifies that by 2020, 90 percent of Near Earth Objects (NEOs) (aka, asteroids) 140 m (150 yards) or more in diameter will be identified, especially those that could possibly impact the Earth. So far, more than 700 potential "planet killers" have been found, out of an estimated 1,000. Fortunately, none are on a collision course with the Earth. Smaller ones, less than 140 m (150 yards) in diameter, are present in much greater numbers (up to 100,000) and are much harder to detect. Of the smaller ones perhaps 20 are "potentially hazardous." To better understand asteroids and the threat they pose, in 2001, the U.S. landed the *Near Earth Asteroid Rendezvous* spacecraft on the asteroid Eros, and Japan landed a probe, *Falcon*, on Itokawa in 2005 to return a sample to Earth. The NEOSSats mentioned in Chapter 14 also will be searching for NEOs.[177]

In the event that a large asteroid was found to be on a collision course with the Earth, there is currently nothing that we could do to stop it. Ideas for how to deflect or destroy such an asteroid range from nuclear missiles to blow it apart (ala the movie *Armageddon*), to impacting it with an unarmed missile to change its course, to using laser beams to destroy it or deflect its course, to placing a spacecraft near it as a "gravity tractor" to gently nudge its orbit. NASA currently has no plans to test any of these possibilities, but the European Space Agency has been working since 2005 on an asteroid-deflection project called the *Don Quijote* mission, that will impact an asteroid with one spacecraft while another nearby observes and monitors the change in path.[178]

Tsiolkovsky envisioned a time when asteroids will be mined for their resources and, in a way, we are already doing that by mining those that have impacted the Earth. The Sudbury nickel-iron asteroid that hit North America 1.85 billion years ago, is one of many that are mined. In addition to copper, iron, and nickel, many of the materials needed to make the motors and batteries for electric vehicles—or the fuel cell stacks for fuel cell-powered vehicles—are rare-earth materials such as platinum, palladium, rhodium, and neodymium that are present in some asteroids. As demand for these elements rises and the cost increases, and as the cost of space access declines, it will become more economical to mine the elements directly from asteroids. But first we need to develop the techniques for mining asteroids.[179]

FRIDAY THE 13TH IS CONSIDERED BY SOME to be an unlucky day, but Friday, April 13, 2029 will bring a rare opportunity. On that day, an asteroid (number 99942, *Apophis*, 2004 MN4) will pass the Earth at a distance of 30,000 km (18,600 miles), which is closer than the geosynchronous satellites at 35,800 km (22,240 miles) above the Earth. *Apophis* is about 320 m (1,000 ft) across, not extremely large, but big enough to do serious damage if it struck the Earth, which it is not expected to do. (Former astronaut Rusty Schweikert has proposed a mission to *Apophis* in 2014 to attach a radio transponder so that the asteroid may be better tracked to more precisely determine the degree of hazard it poses.) But *Apophis* is also big enough to serve as a test bed for mining asteroids. It will be traveling at a very high speed, but, if it is not rotating too rapidly, a robotic mission could intercept it and begin automated mining operations. *Apophis'* orbit will be changed by the close encounter with the Earth, but it is thought possible that it will again pass near the Earth in 2035, though not as closely. If it does return to the vicinity of Earth, at that time the products of the mining operation could be retrieved. By 2029, the technologies certainly could be developed to perform this mission (and even by 2014 some rudimentary methods could be developed for testing, with retrieval of products in 2029), and it would be a way to validate techniques for mining the asteroid belt. But before mining the asteroids (and the smaller meteoroids) in space, we could mine those that have impacted the Moon.[180]

One goal of NASA's Constellation Program was to establish a permanent human presence on the Moon, which would include developing mining techniques. The surface of the Moon is scarred by countless impacts—most from meteoroids that would have burned up before reaching the surface if the Moon had an atmosphere like the Earth. These meteoroids contain the materials needed to construct buildings and other structures on the Moon and could be used to build structures in space. They also contain the rare-earth elements mentioned earlier, which could be refined and shipped to Earth.

Other resources that might be mined from Lunar soil include helium-3 and water. He-3 could be extremely valuable for fusion power plants because it does not make the reactor radioactive as other fusion fuels do. In 2009, NASA's *Lunar CRater Observation and Sensing Satellite* (*LCROSS*) mission to the Moon found significant water ice by impacting a crater near the south pole. Mining to retrieve water ice could be even more important, as a source of drinking water and breathing oxygen, and, more critically, it could be used to make rocket fuel.[181]

The techniques to mine asteroids and the Moon need to be developed, and government support in the form of research funding and missions to test techniques is essential. A mission to *Apophis* would serve to focus efforts and advance the state-of-the-art. Spinoffs could include less environmentally destructive methods for mining on the Earth.

Proposal 3: Promote the acquisition and utilization of the resources in space by performing research and technology demonstration missions, on the Moon and to asteroids such as *Apophis*.

16

Manufacturing in Space

"[Tracking and cataloguing the asteroids that might hit Earth] leads naturally to unmanned reconnaissance missions, culminating eventually in space mining and manufacturing." — Senator Spark Matsunaga

MANUFACTURING IN SPACE AS A COMMERCIAL ENTERPRISE was a stated goal for the Space Station *Freedom* program, and initial efforts have been performed on some space missions. One quality that makes the space environment different is what used to be called zero-gravity and is now usually called micro-gravity—the lack of sensation of gravitational forces.[lxi] Other characteristics of space that may be important for manufacturing include the intense, constant solar energy; the coldness when the sunlight is blocked; the hard vacuum; and the space itself, enabling the manufacture of very large structures.

• • •

WHEN PRESIDENT REAGAN INAUGURATED Space Station *Freedom*, commercial production of marketable products was expected to be performed on board. The micro-gravity condition allows for growing purer, defect-free crystals of various compounds, mixing alloys of compounds that will not mix on the Earth, and making precisely sized microspheres. Any of the special characteristics of being in space could be used to make products

---

[lxi] Gravity is present throughout space, of course, but a spacecraft in orbit around the Earth or moving through space without propulsion is in free fall and so the sensation of being in a gravity field is not felt.
A more technical definition of micro-gravity is: the condition of apparent weightlessness that occurs when the centrifugal force and the gravitational force acting on a body exactly counterbalance.

that today cannot be made on the Earth. Numerous experiments on board the Space Shuttles, *Spacelab*, and the *ISS* involve processing pharmaceuticals to make purer drugs, for example, or growing extremely high quality crystals of various materials to better understand how they form or to use them to improve various devices that rely on such crystals. Pharmaceuticals processing was the first real manufacturing in space, onboard the Space Shuttle *Discovery* on August 30, 1984, by Charles Walker, a McDonnell Douglas Astronautics Company engineer and scientist. These experiments led to improved methods of production on Earth. The first commercial products made in space, however, were 10-micron latex spheres used for calibrating microscopes. At the other end of the scale from microspheres, pharmaceuticals, and crystal growth, micro-gravity also enables the construction of extremely large structures.[182]

ON EARTH, THE TALLEST BUILT STRUCTURES are skyscrapers, several of which are over 400 m (a quarter mile) in height. In the United Arab Emirates, the Burj Khalifa (called the Burj Dubai before opening in January 2010), has far exceeded those at 828 m (more than ½ mile), at a cost of $4.1 billion. The longest structures are bridges,[lxii] some of which have total spans several km (miles) long, and the largest clear-span structure is The O_2 (formerly the Millennium Dome) in London, England with a span of 365 m (1,200 feet). The largest free-floating structures currently being made are supertankers for transporting petroleum. The largest of these are over 400 m (1,312 ft) in length. All are impressively large, but someday much larger structures may be built.[183]

At the Montreal Expo '67, the U.S. pavilion was a geodesic ¾ sphere, 76 m (250 ft) in diameter, designed by Buckminster Fuller. It was a spectacular structure, with a monorail train running through it, that symbolized the forward-looking theme of the Expo. Fuller envisioned much larger domes to cover entire cities,

---

[lxii] Actually, the Great Wall of China, at about 6,400 km (4,000 mi), could be considered the longest built structure, or the U.S. Interstate highway system, with a total length of 75,440 km (46,876 mi).

protecting them from harsh weather and increasing their energy efficiency, and made a sketch showing much of Manhattan, New York, covered by such a dome. Constructing it would be a tremendous challenge, but the idea is certainly intriguing.[184]

In 1958, Fuller proposed a dramatic variation when he realized that a geodesic sphere larger than 800 m (about ½ mile) in diameter would be able to float in the atmosphere, if the air inside was warmed by only one degree. Just as a hot air balloon rises because the heated air is less dense than the ambient air, even when made of steel, a large enough sphere would rise when the air inside is warmed. Depending upon the temperature, the moisture content in the atmosphere, and the size of the sphere, the buoyant force can amount to hundreds of thousands of tons with only a few degrees temperature increase. There could even be openings in the sphere since the pressure would be equalized. Geodesic spheres make such efficient use of material that the mass of the structure would be more than offset by the reduced mass of the air within, due simply to the reduction in density from the warmer temperature.[185]

Such a sphere could support a community and be a floating city in the sky or, as Fuller labeled the idea: a Cloud Nine. The heat to warm the air could come from heating by the sun or from normal human activities. A sphere 1.6 km (one mile) in diameter could support a community of several thousand people. Cloud Nines could be the epitome of environmentally responsible living, not requiring the use of any land or involving the destruction of habitat or the paving of any farmland. A Cloud Nine would be a very energy efficient place to live, too, and would be like living in a small town with everything within walking distance. It could be tethered to the ground or moved about to visit different places or to avoid poor weather. (A Cloud Nine could also be used in an emergency to respond rapidly to natural disasters, providing food, medical care, and shelter.) A sky elevator could provide access, or small aircraft or skycars could be used to come and go. Such a Cloud Nine would be a highly efficient use of materials and energy, and the advantages warrant consideration of how one could be constructed efficiently. The scale is not really much different from that of other planned cities that have been built or are under construction (e.g., Masdar City). The challenge would be fabricating such a structure, that would enclose a much greater volume than anything ever constructed. Lifting the construction materials up to 1.6 km (a mile) into the sky—even when the materials are high-strength and lightweight—would require considerable energy (involving the burning of much fossil fuel using current methods of construction). Could such a geodesic sphere be manufactured in space?[186]

In space there are fewer constraints on the size to which a structure can be built. The constraints relate to the materials of construction and the construction process. If, for a large structure, the materials of construction were produced on the Earth and launched into space for assembly by astronaut construction workers, such a project would be highly impractical. But materials are already in space. Asteroids are one source, but they are moving rapidly, so the Moon may be the best source. With 1/6 the gravity of Earth and no atmospheric friction to overcome, lifting materials from the Moon is much easier than lifting them from the Earth. The Moon also has abundant solar energy for processing materials mined from its surface into a useable form. One advantage of a geodesic-type structure is that individual parts are relatively small, so after they are fabricated they could be delivered by a Lunar elevator or launched into space with a mass driver (an electromagnetic rail that can very rapidly accelerate objects to orbital velocity[lxiii]). They would then be snagged at the construction site, perhaps the L1 Lagrange point between the Earth and the Moon. Specially designed construction robots could then assemble the sphere, piece by piece. A team of robots could build one as quickly as parts are delivered.

After a sphere is completed and roughly outfitted, the next challenge would be getting it to the Earth. The total mass of a

---

[lxiii] The mass driver was invented by Gerard K. O'Neill in the 1970s. The Space Studies Institute, http://ssi.org/about/history/,
http://web.archive.org/web/20080113061809/http://ssi.org/?page_id=5,
http://www.astro.virginia.edu/~jkm9n/teaching/astr342/lectures/July2.pdf

Cloud Nine shell might be millions of kilograms (pounds), but since it does not have to overcome the Earth's gravity, a "nudge" in the proper direction would be sufficient to begin its descent toward the Earth. A solid object of that size falling to Earth would cause massive destruction. The Cloud Nine shell, however, would be far from solid, having an average density not much greater than the vacuum in which it was constructed. By way of comparison, consider the difference between a bowling ball and a basketball being thrown into a pool. The bowling ball makes a huge splash, plowing through the water to the bottom, whereas the basketball makes a much smaller splash, goes only a short way into the water (the depth depending on its velocity) and then bobs to the surface with some portion above the water. A Cloud Nine entering the atmosphere would behave much more like the basketball than the bowling ball. As it hits the atmosphere it would be traveling at a high speed, but it would decelerate extremely quickly due to its low density. There would be some heating, so the structure would have to be designed to withstand the heat, but not nearly so much as a Space Shuttle returning to Earth, which is closer to a bowling ball in density. The heating would actually add to the sphere's buoyancy so that it would remain high in the sky, gently settling as it cooled. It could then be tethered to the ground for final outfitting, much as the internal systems in skyscrapers are installed after the structure and shell are built.

Whether or not this is something that actually can be done remains to be seen. There are a number of complications, from all of the spacecraft orbiting the Earth that a descending Cloud Nine would have to dodge, to ensuring stability of a completed Cloud Nine when winds blow from different directions at different altitudes, much as cruise liners must be actively stabilized to keep them from tipping over due to ocean waves. But a Cloud Nine is an example of something that might be manufactured in space more easily than on the Earth.

OF COURSE, IT WOULD ALSO BE POSSIBLE to build a Cloud Nine and leave it in space. Large space cities orbiting Earth or in independent orbits around the sun have been proposed by many people. In 1976, Gerard K. O'Neill, proposed space cities large enough to have soil with plants, lakes, and clouds inside. Those would be cylindrical and would spin to simulate gravity, although others could be spherical or non-spinning. In 2005, Buzz Aldrin (the second person to walk on the Moon) proposed placing space cities in eccentric orbits that go between the Earth and Mars. He calls them Mars Cyclers, because they would "cycle" between Earth and Mars. They could even go out to the asteroid belt to pick up raw materials. Such orbiting cities, once in place, would reduce the cost to bring mineral resources from the asteroid belt to Earth. By outfitting a Cycler as a factory, during the year or so transit time, the raw materials could be processed into manufactured products, similar to the way that Ford operated the River Rouge plant in Detroit, in which raw materials entered one end and finished automobiles rolled out the other. With the resources of the asteroid belt, energy from the sun, and the ability to utilize micro-gravity, new types of high-strength, lightweight products would be possible, and new industries could be created.[187]

A number of technological advances are necessary in order for Cloud Nines to be constructed in space, including techniques for mining the Moon or asteroids (mentioned in the previous chapter), processing the raw ores, developing a full-scale mass driver or Lunar space elevator, and advancing robotics for construction in space. With a few key technologies, developed and tested by 2020, Cloud Nines could be manufactured, perhaps by 2030.[188]

In the near term, manufacturing of other products that cannot be made on the Earth is possible on the *ISS* or privately developed space stations. For the longer term, for manufacturing that unavoidably involves the use or creation of hazardous materials, manufacturing in space would be preferable.[lxiv] With the right

---

[lxiv] For example, the manufacture of flat panel monitors, television screens, and thin-film photovoltaic panels uses nitrogen trifluoride as a cleaning agent. Although present in the atmosphere at the parts per trillion level, NF_3 is rapidly increasing and it is a greenhouse gas that is thousands of times more potent than CO_2. Borenstein, Seth, "2 greenhouse gases on the rise worry scientists," Associated Press, October 24, 2008, http://www.usatoday.com/tech/science/environment/2008-10-24-greenhouse-gases_N.htm.

incentives, products manufactured in space could become more common. One idea for an incentive is called "zero-G, zero-tax," which would eliminate taxes for space-related activities and products made in space.

Proposal 4: Promote manufacturing in space—from micro- to macro-scale products—through tax incentives, direct funding, and by the government serving as a pioneer customer. Promote the development and demonstration of the robotics and techniques that could be used to construct Cloud Nines (and perform other activities in space), and support an in-space demonstration of a Cloud Nine assembly and entry into the atmosphere with a small-scale version.

17

Power from Space

"I think we need to take positive steps toward space solar-power systems. We need to move in a step-by-step manner. It's a real possibility to have a great new energy source for mankind." — Dana Rohrabacher, March 5, 2001

TO COUNTER GLOBAL WARMING, SOME PROPOSE building a huge shield in space to block a small percentage of the radiation from the sun.[lxv] This is an intriguing idea that may indeed be effective, but it would deal only with one effect rather than the cause of global warming and would reduce the output of solar power plants on Earth, as well as have unknown effects on the biosphere. It would also be very expensive to implement, although by utilizing material from the Moon to build the shield the cost could be lower than by using material launched from the Earth. Addressing the causes or contributing factors of global warming directly would be more effective and have a lower overall cost. Solar power satellites would do this by providing electricity to offset fossil-fuel sources.

• • •

CONVERSION OF SUNLIGHT TO ELECTRICITY was first achieved in 1883 by Charles Fritts, though it was only a laboratory curiosity for many years. In the 1960s, NASA developed photovoltaic (PV) solar cells for use on spacecraft, to avoid the

---

[lxv] Edward Teller and James Early proposed building a 2,000 km^2 shield at the Earth-sun L-1 point to block a few percent of solar radiation. Simon P. (Pete) Worden and Roger Angel propose using millions of smaller "blocker spacecraft" for the same purpose. This would require 10 to 20 million tons of material. Worden, Simon P., "The Vision for Space Exploration: New Opportunities," speech at the International Space Development Conference, May 7, 2006, http://www.spaceref.com/news/viewnews.html?id=1119

limitations of batteries. Today, almost all spacecraft are powered
by converting sunlight into electricity (except for a few deep-space
probes that require nuclear sources due to their distance from the
sun and a few short-duration spacecraft that rely on batteries).
Since the 1960s, the conversion efficiencies of PV power have
increased to as much as 40 percent for rigid solar cells and up to 30
percent for flexible sheet PV material. Efforts to further increase
the efficiency of PV solar cells are ongoing, as well as efforts to
improve manufacturing techniques.[189]

In 1968, Peter Glaser proposed having solar power arrays in
orbit about the Earth. His idea was to beam the generated power as
microwaves to receivers on the Earth. At the distance of Earth's
orbit from the sun, the intensity of solar energy in space is 1,366
W/m^2 (referred to as the "solar constant" although there is some
variation). This includes the entire electromagnetic spectrum, not
just visible light, and is much greater than the intensity of solar
energy at the Earth's surface, which, for the latitude range of North
America, is 125 to 375 W/m^2.[190]

Several studies since the 1960s have concluded that space
based solar power (SBSP) is technically feasible, but that "the
prohibitive cost of lifting thousands of tons of equipment into
space makes it uneconomical." Recent advances in lightweight PV
materials and increasing efficiencies in energy conversion are
reducing the mass that would need to be launched into orbit, and
proponents (including the National Space Society, the Space Solar
Alliance for Future Energy, and the Space Power Association) are
pressing to develop SBSP systems. Japan is planning to have an
SBSP system in operation by 2040.[191]

In 2007, Kevin Reed proposed a demonstration project to build
an 80 m (260 ft) rectifying antenna in Palua (in the South Pacific
Ocean) to convert microwaves beamed from a PV array orbiting
480 km (300 mi) above the Earth into 1 MW of electricity. The
SBSP satellites would pass over once every 90 minutes, allowing
power to be beamed for about 5 minutes. Orbiting that close to the
Earth, SBSP satellites would be in the Earth's shadow for about
half of each orbit. By placing them in higher orbits the duration in
sunlight would increase, which would increase their effective

efficiency. At a sufficiently high orbit, sunlight is available continuously, without the intermittency of a day/night cycle, with only brief periodic eclipses. If placed in GEO, SBSP satellites will remain above a particular location on Earth and can provide power continuously.[192]

If ground-based PV arrays provided all the electricity in the world, as many as 220,000 sq km (85,000 sq mi) would be needed. Many could be placed on roof tops, but PV power plants have also been placed in deserts. Most deserts, though, are active ecosystems that would be affected by installing a PV power plant. Potential new large uses of electricity, such as electrolyzing water to generate H_2 and desalinating seawater to provide fresh water, would increase the area needed. One advantage to having a PV array in space is that, due to the greater intensity of solar energy, the PV area required would be far less, even accounting for transmission and conversion losses. Another advantage is that no ecosystems would need to be disturbed.[lxvi]

Placing PV arrays at GEO would provide some interesting possibilities. For one, rather than beaming energy to a location on the Earth, it may be possible to physically connect the array using a conductive carbon nanotube cable. There would be transmission losses (unless a superconducting cable was developed), but a sufficiently conductive cable might be preferable to beaming power to Earth. Additional PV arrays could, eventually, make a

---

[lxvi] The area of PV arrays required on Earth and in space that would be necessary to produce the 15.883 TWhr of electricity generated worldwide in 2003, can be estimated by making some assumptions about availability, conversion efficiency, and transmission efficiency. For comparison purposes, let's assume the availability would be 40% on Earth (to account for night and clouds) and 80% in space (to account for portions not facing the sun and periods in shadow), with a conversion efficiency of 25% (on Earth and in space), and transmission efficiency of 90% on Earth and 30% from space. For an average energy intensity of 0.2 kW/m^2 on Earth and 1.3 kW/m^2 in space, the area required on Earth would be 100,550 km^2 (38,800 mi^2), whereas the same amount of electricity could be provided by 23,229 km^2 (9,000 mi^2) of PV arrays in space. Actual generation and loss percentages would likely be somewhat different, but these calculations indicate that a far smaller area of PV arrays in space could provide the same amount of electricity.

ring completely around the Earth, like a thread across the sky. At GEO, the radius of the orbit (from the center of the Earth) is about 42,000 km (26,000 mi) for an orbital length of about 247,000 km (163,000 mi). Therefore, a ribbon of solar cells 1 km (2/3 mi) wide that completely encircles the Earth could provide many times the electricity currently used. And it would generate electricity every hour of every day.

According to a 2007 report prepared for the National Security Space Office, several major challenges will need to be overcome to make SBSP a reality, including the creation of low-cost space access and a supporting infrastructure system on Earth and in space. The government will need to perform three major tasks to catalyze SBSP development. The first is to reduce a major portion of the early technical risks by an incremental research and development program resulting in a space-borne proof-of-concept demonstration. The second challenge is to facilitate the policy, regulatory, legal, and organizational instruments needed to create the partnerships and relationships (commercial-to-commercial, government-to-commercial, and government-to-government) needed for this concept to succeed. The final government contribution is to become a direct early adopter and to provide incentives for other early adopters similar to incentives for energy producers today.[193]

Proposal 5: Advance the development of space-based solar power systems through research and a concept demonstration program, with the goal of having a validated demonstration SBSP array operational by 2020.

18

Technology to Reach Space

"Mankind will not forever remain on Earth, but in the pursuit of light and space will first timidly emerge from the bounds of the atmosphere and then advance until he has conquered the whole circumsolar space." — Konstantin Tsiolkovsky

THE HIGH COST OF ACCESS TO SPACE is the number one challenge to space activities. Proposal 2 addresses this regarding robotic spacecraft, but sending people to space is even more challenging. In February 2010, President Obama proposed supporting commercial access to space by using privately developed rockets to carry astronauts to and from the *ISS*. Such commercial capability will be a major step toward greater utilization of space. As presented in Chapter 9, to fully utilize the potential of space activities, however, a means of getting to space other than by using rockets will be needed, that does not have the environmental issues of rockets and that could greatly reduce the cost of access.[194]

• • •

REDUCING THE COST OF ACCESS TO SPACE is essential. After 50 years of using rockets, they are still too costly, and even with reductions by mass production, there would be environmental costs to launching rockets as frequently as needed for major utilization of space. An alternative is needed, and the space elevator concept is among the most promising.

The first space elevators will have to connect with the Earth at, or very near, the equator. While platforms floating in the ocean may be suitable, it may be that equatorial mountains provide better attachment points, at least initially, so those equatorial countries with tall mountains will have something that other countries would want access to.

Locations in several countries on or within a few degrees of the equator could serve as ground stations, some on mountains, others not: Quito, Equador; Libreville, Gabon; Nairobi, Kenya; Kiribati; Indonesia; Singapore. As improvements in cable strength are made, the ground station could be located tens of degrees off of the equator (requiring longer cables), and many more cities would be able to serve as ground stations. Suitably pairing elevator stations north and south of the equator that attach to the same terminal at GEO would balance the tension on the cables so the space terminal would remain in place. Forty or so degrees north or south latitude may be a practical limit, but such pairs could include New York City, NY (40° 42' N, 73° 58' W) and Puerto Montt, Chile (41° 28' S, 72° 56' W); Lisbon, Portugal (38° 42' N, 9° 11' W) and Tristan da Cunha Island (37° 15' S, 12° 30' W); near Tijuana, Mexico (32° 32' N, 117° 1' W) and Easter Island (27°09'S, 109°27'W); Tokyo, Japan (35° 40' N, 139° 45' W) and Canberra, Australia (35° 30' S, 149° 0' W); Athens, Greece (37°58'N, 23°13'E) and Cape Town, South Africa (33 °56'S, 18 °29'E); Hong Kong, China (22°15' N, 114°11' E) and Carnarvon, Australia (24° 52' S, 113° 38' E); or Taipei, Taiwan (25° 1' N, 121° 22' E) and Brisbane, Australia (27°29' S, 153°8' E).[195]

Eventually, by further extending this idea, ground stations could be even more distributed with multiple elevators—north and south of the equator to balance the forces on the terminal—connected to a single terminal at GEO. A few terminals distributed around the Earth could provide direct access to space for dozens of locations on the ground. Such grouped locations might include:

Atlantic Ocean Node: Quito, Equador; New York City, U.S.; Miami, U.S.; Havana, Cuba; Rio de Janeiro, Brazil; Lagos, Nigeria; Buenos Aires, Argentina; Cape Town, South Africa; Rome, Italy; Paris, France

Indian Ocean Node: Singapore; Perth, Australia; New Delhi, India; Antalaha, Madagascar; Ankara, Turkey; Hong Kong, China; Beijing, China; Dubai, UAE; Karachi, Pakistan; Tehran, Iran

Pacific Ocean West Node: Tokyo, Japan; Honolulu, U.S.; Sydney, Australia; Singapore; Aukland, New Zealand; Taipei, Taiwan

Pacific Ocean East Node: Honolulu, U.S.; Los Angeles, U.S.; Mexico City, Mexico; Quito, Equador; Santiago, Chile; Lima, Peru; Micronesia

With such a distribution of space elevator terminals at GEO, access to space would be readily available. With a solar array ring connecting them, providing power for the climbers as well as producing power for Earth, this could be the beginning of another concept of Arthur C. Clarke: a "Ring City" habitable structure. Clarke envisioned, in the distant future, that communications satellites would not need to be launched to particular locations. Instead, communications equipment could be lifted with the space elevator and, using an "internal railroad," positioned along a track or tether at specific locations in GEO.[196]

Especially as more space elevators are built, debris and unused spacecraft in Earth orbit will be an increasing concern. A means of clearing that debris will need to be developed, to minimize damage to the cables by high-speed collisions.[197]

> **Proposal 6**: Advance the development of a space elevator and related technologies through funding of research and by prizes or similar awards for key developments, such as by expanding the current Centennial Challenges program. The goal would be a demonstration sub-scale elevator at GEO (or a Lunar elevator) by 2020.

19

International Agreements

"We believe that when men reach beyond this planet, they should leave their national differences behind them." — President John F. Kennedy, 21 February 1962

"Space can be explored and mastered without feeding the fires of war, without repeating the mistakes that man has made in extending his writ around this globe of ours. There is no strife, no prejudice, no national conflict in outer space as yet. Its hazards are hostile to us all. Its conquest deserves the best of all mankind, and its opportunity for peaceful cooperation may never come again." — President John F. Kennedy

DEVELOPING THE TECHNOLOGIES is one challenge of becoming a space-faring civilization, minimizing political and social conflicts is another. International cooperation and agreements will be required on issues that arise in order to avoid a "wild west" phase of development. These will especially relate to standards, property rights, and weapons in space.

• • •

TO MINIMIZE CONFLICTS and to promote cooperation, international agreements on the utilization of space will be essential. Having well-defined standards would be especially critical in emergency situations to allow rescue operations and related actions. Standards related to communications and power systems, atmosphere condition for habitats, interfaces for docking ports and resupply connections, and other considerations are being defined by the ISECG, as mentioned in Chapter 11. All space-faring countries will benefit by adopting such standards.

Because of the different circumstances, laws we take for granted on Earth may not apply in space, so special conflict rules may need to be developed. Current treaties dealing with use of the oceans (e.g., the Law of the Sea Treaty) and Antarctica may serve as starting points, but they are inadequate when considering space. The Outer Space Treaty of 1967 primarily addressed weapons in space and is inadequate when considering the full range of potential activities. One challenging issue relates to ownership of any materials mined in space or of any bodies in space (the Moon, Mars, asteroids, etc.). International agreements in other areas will also be necessary, especially concerning regulation of commerce, commercial development, property rights in general, criminal activity, and personal interactions. The ISECG's efforts to establish standards could be extended to address issues relating to property rights and the use of materials found in space. Any such agreements should also establish preserves to be off-limits to settlement or extraction of resources, to protect places such as the *Apollo* and *Surveyor* landing sites on the Moon; landing sites of spacecraft on Mars, Venus, and other locations; Europa, a moon of Jupiter; Titan, a moon of Saturn; and other locations found to have liquid water, potentially harboring life; and other unique locations in the solar system.[198]

Regarding weapons in space, because of the high velocities involved, any spacecraft could be used to destroy another spacecraft. Countermeasures include tracking of spacecraft in Earth orbit and designing spacecraft to have maneuvering capabilities. This is a situation where "a miss is as good as a mile" and the extreme precision needed makes this an unlikely threat. Of greater concern is the use of missiles that could be directed to attack spacecraft. In 2008, China destroyed one of its aging weather satellites orbiting 864 km (537 miles) above the Earth as a demonstration of this capability. One result was generation of debris that increases risk for other spacecraft, including the *ISS* and China's own spacecraft. Of even greater concern is the possibility of stationing weapons, including nuclear weapons, in space. No country is known to currently have such capability, but ensuring that remains the case will require vigilance and international cooperation of the highest degree. Global Zero, an international group that has broad support from "political, military, business, religious, and civic leaders" seeks "to eliminat(e) nuclear weapons over the next 25 years." They hope to hold a world summit "to encourage leaders to meet to discuss and eventually negotiate a timetable for disarmament." Such efforts can help ensure that no weapons are used in space as well. Soon after being sworn in as president in January 2009, President Obama proposed a ban on weapons in space. The complexities associated with such efforts are described by Michael Listner and include defining just what a space weapon is and enforcing an agreement that bans them. This is not a trivial problem, but it is not intractable either.[199]

International cooperation and coordination is essential. All of the experience and skills acquired during previous negotiations— such as the bilateral negotiations for the Camp David Accords which led to the 1979 Israel-Egypt Peace Treaty and the Strategic Arms Reduction Treaty (START 1) initiated in 1982 to reduce the nuclear weapons arsenals of the U.S. and the USSR, as well as multi-national efforts such as the Montreal Protocol phasing out ozone-destroying chemicals—should be called upon in developing the rules for our activities in space.[200]

Providing the legal framework to support the development and settlement of space is essential to minimize the occurrence and severity of conflicts, which will inevitably arise. An international joint venture along the lines of INTELSAT, with many countries participating, may serve as a model that would minimize conflicts. Providing the opportunity for countries—and companies and individuals, even—to participate to the extent that they are willing and able would broaden the opportunities and the rewards, while distributing risks. As Greg Anderson states in a 2008 article in *The Space Review*: "Space (development) must be ... good for society as a whole." A joint venture approach would also broaden the support for space activities and would help ensure that those activities are good for society. A suitable legal framework, that serves as a Space Mining Act or Space Homestead Act, would encourage private investment and promote peaceful development and settlement of space.[201]

Proposal 7: Negotiate international treaties and agreements relating to property rights and development of space, including agreements that serve the purposes of the Homestead Act and Mining Act for the American West (with provisions to protect sensitive locales and require responsible development). Create an organization similar to INTELSAT to perform specific space development activities.

Part V

Roles for All

"... in a very real sense, it will not be one man going to the moon—if we make this judgment affirmatively, it will be an entire nation. For all of us must work to put him there." — President John F. Kennedy

• • •

ADDRESSING THE CHALLENGES of the 21st century requires long-range consideration and near-term action. Due to the magnitude of the challenges and the breadth of effort required to address them, cooperation will be needed on a grand scale, for space activities as well as for energy and environmental issues. Through coordinated action, we each can have a role in successfully resolving the challenges, and one goal should be to enhance opportunities to participate—for individuals, organizations, institutions, and businesses, small and large.

The government and private enterprise have distinctly different roles, but ideally they are complementary and mutually supportive. One fundamental difference is the time scale of consideration—the length of near-term and long-range planning and projects. For the majority of businesses, planning is for the next fiscal quarter to, perhaps, three years, due to uncertainties in forecasting the marketplace. For government, in contrast, long-range planning often extends to 30 years or more. Another fundamental difference between government and private enterprise is that the government must consider a broad range of factors, whereas private enterprise has more narrowly focused interests. The roles of academia—education and research—are essential to ensure a knowledgeable public and to advance our understanding. Individuals as the initiators of ideas and advocacy groups as the voices of constituencies, have vital roles in bringing attention to issues and providing support for particular positions.

20

Government Roles: Vision and Policies

"In outer space you develop an instant global consciousness, a people orientation, an intense dissatisfaction with the state of the world, and a compulsion to do something about it. From out there on the moon, international politics look so petty. You want to grab a politician by the scruff of the neck and drag him a quarter of a million miles out and say, 'Look at that, you son of a bitch.'" — Edgar Mitchell, Apollo 14 astronaut

FOR SUCH LARGE-SCALE DEVELOPMENTS as the interstate highway system, the Tennessee Valley Authority and rural electrification, the national park system, the air traffic control system, and the Apollo program, one government role was establishment of the vision to be achieved. The initial idea was conceived by an individual and usually championed by a group, but to become reality the government had to establish the vision as a societal goal and create the policies needed to achieve it.

Where large-scale or long-term actions are required, the government's role is essential. Where international agreements are important, the government's role is clear. Where specific actions by companies or individuals are desired, but initial investment requirements are high, the government also has a role. In general, the government's roles are to establish a vision and provide stability for the achievement of that vision. This requires taking a long-term view as expressed through actions such as formulating policies that support the vision, negotiating any international agreements that are needed, clarifying boundaries to reduce the risks of striving for the vision, performing essential tasks that are beyond the capabilities of private enterprise to perform, and promoting a stable business environment that allows those who so desire to participate.

As happened with the Apollo program, the next phase of space activities will require the government to take a leadership role, but in a manner that supports independent commercial activities, with the aim of further developing space-related industries that will not rely on government support.

••••

THE GOVERNMENT'S ROLE REGARDING SPACE ACTIVITIES is to prepare for and enable the coming convergence and rapid growth phase. Given the issues that society is facing, the technology development activities already underway, and the efforts of entrepreneurs to develop space access and operations, the convergence that will enable rapid progress in space is imminent. Clear, forward-looking policies are needed to provide the guidance for government and private enterprise activities.

The vision presented in this book is one of abundance—addressing energy and resource acquisition, environmental preservation, and international cooperation. A broad program of space-related activities will help to achieve these goals—from fundamental research and applied development, through detailed exploration of the solar system, to promoting space-based industries. The participation of companies and individuals is essential, but the government's role will be key to coordinating actions and reducing the risks for participation. The government (through NASA, DOE, and other agencies where appropriate), as provider of the institutional backing for this vision, should immediately:

- Establish an International Space Decade, defining and coordinating the missions—the specific destinations, the types of instruments needed, the communications and control considerations, coordination with foreign countries, etc.

- Establish the 1,000 spacecraft program and provide guidelines for the minimum capabilities of the spacecraft. Also, promote sharing among spacecraft manufacturers, of the information and technologies needed to produce the most capable. spacecraft.

- Enhance current efforts to promote development of commercial space capabilities. With the 2011 budget proposal from the Obama administration, NASA will do this by "providing industry with NASA technical expertise," "serious seed money ... and a firm commitment to buy crew transportation services."[202]

- Accelerate research and development of the space-related technologies that directly address energy and environmental concerns and that will be needed for major utilization of space—carbon nanotube cables, radiation countermeasures, miniaturization of spacecraft components, propellantless propulsion, asteroid mining techniques, power beaming, magnetic rail, PV arrays in space, etc.

- Initiate international negotiations for space utilization agreements (including property rights, weapons control, policies for power beaming, and other legal aspects) and to establish an INTELSAT-type organization to begin utilization activities.

- Initiate preparations for in-space validation—by 2020—of space elevator construction techniques, space-based solar power satellite operation, and Cloud Nine assembly and entry.

- Promote rapid spin-off of technical advances made for space application into products for earth-bound use.

- Implement risk-reduction efforts to promote private enterprise participation, through a combination of direct funding of research and development (R&D) efforts, offering prizes for specific results, and tax incentives.

SEVERAL TYPES OF RISKS that are routinely faced by private enterprise can be reduced by appropriate government actions. Technological risks can be reduced via government-sponsored research, as was done by the NACA to advance aeronautics, or by providing tax incentives for particular research activities. Financial risks can be reduced via direct funding, subsidies, or tax breaks, as is done today regarding energy acquisition and use. Marketplace risks can be reduced by the government serving as a "pioneer customer" to establish a market, as was done to encourage aviation. Establishing clear boundaries also reduces risks, either geographic boundaries, as for settlement of the American West, or legal boundaries that may involve international agreements, as with the Montreal Protocol to replace chlorinated fluorocarbon (CFC) compounds—used in refrigerators, air conditioners, fire extinguishers, and other products—with more advanced and environmentally benign chemicals. Technology development and commercialization can also be promoted by direct government funding. The overall goal of these efforts is to have—by 2020—the understanding of how space activities can directly address societal issues and the means to carry out those activities.

Providing educational opportunities is also key to ensuring that the skills and knowledge needed to address our challenges are available when we need them, and another role for the government is to promote quality education. The government can, and does, promote education in several ways, through student aid, funding of schools, and funding programs at educational institutions, including NASA's NSCORT[lxvii] program. By such actions the government can ensure that students acquire the best education possible and that research is performed in the areas of interest. Given the importance of education, even more needs to be done to ensure that the knowledge, skills, and experience needed in the 21st century are broadly available.   .

[lxvii] NASA Specialized Center of Research and Training

166

21

Education and Academia: Discovery

"Every generation has the obligation to free men's minds for a look at new worlds ... to look out from a higher plateau than the last generation." — Ellison S. Onizuka, astronaut and the first Asian American in space

"The beginning of knowledge is the discovery of something we do not understand." — Frank Herbert

"What the space program needs is more English majors." — Michael Collins, Apollo astronaut

ADDRESSING THE CHALLENGES OF THE 21ST CENTURY will require our best efforts, and, thus, will require knowledge of the key fields and an educated workforce to use that knowledge. Where knowledge was hard-won with years of effort by many people, it is essential that it be transferred to the next generation so they have a foundation upon which to build further advances and find solutions to the challenges they face. The understanding necessary to assess "the physics" of proposed solutions, as well as the environmental and social impacts, will be key to successfully implementing them.

Fundamental research, in contrast to product development, is typically performed at academic institutions whose "product" is knowledge, although corporations do perform some fundamental research, especially under government sponsorship. Much of the long-term research funded by the government is performed at universities. Such research is key for the discovery of new understanding, and developing and testing new ideas.

Learning

T HE ENGINEERS AND SCIENTISTS WHO GOT US TO THE MOON with Apollo were mostly in their 20s and 30s, fulfilling a vision presented by a 43-year-old president. At the time of the *Apollo 11* landing on the Moon, the average age of the NASA workforce was 26. They were young, just embarking on their careers, and proud to have such an important assignment, responding to the challenge of the USSR. By 1990, however, NASA found itself facing a dilemma. After many years of limited hiring, those same people had shifted the age distribution of the workforce and the average age was 41 and climbing. The bell curve of workforce age was inverted and weighted toward the right. Much of the knowledge gained through years of experience on space projects would be lost as the number of retiring employees increased dramatically in the coming years. To retain this knowledge, it was decided that it should be documented to preserve the "cultural heritage" of how NASA accomplishes its missions and the lessons learned over the years. Each technical area of the organization was to prepare a report describing how its tasks were performed.[203]

Life support was one critical technical area to be documented and, at the weekly ECLS group meeting on November 14, 1990, I was surprised to be assigned this task. I had been in the ECLS Group for about four years and had worked for NASA for seven years so I was not exactly a newcomer, but I was hardly an "old timer" either. As a co-op in the Design Integration Branch I had researched life support technologies to evaluate configurations that would make effective systems and I had written a 20-page paper on spacecraft life support systems for an Advanced Ecology class. I don't believe my boss knew about that paper, but it did provide a good foundation for the ECLS design book. After three years of effort, with input from many engineers who had worked on life support systems far longer than I had, *Designing for Human Presence in Space: An Introduction to Environmental Control and Life Support Systems* was published as a NASA Reference Publication (NASA RP-1324). That report serves its purpose well by documenting in one volume the process used to design ECLS systems. It contains technical information, as well as descriptions of the analysis methods and historical spacecraft ECLS systems, along with the lessons learned over the years. Especially after it was available on-line, it was used not just by NASA engineers, but also by teachers, students, and researchers from around the world.

THE ISSUE OF WORKFORCE CHANGE and loss of institutional knowledge is again a great concern, not just for NASA but for the entire aerospace industry. As reported by Dave Montgomery in 2008, "roughly a quarter of the nation's 637,000 aerospace workers could be eligible for retirement ... raising fears that America could be facing a serious skills shortage in the [aerospace industry]." There is also concern that "colleges and universities are turning out far too few engineering and aeronautical graduates to fill future vacancies," and "many young job-seekers ... now regard aerospace plants as 'old-fashioned industries'." But the Generation X and Millennial Generation students who do choose engineering want to do great things that address the issues of concern to them. They are seeking purpose and meaning for their lives, and ensuring the sustainability of the Earth is of growing importance to many of them. They want good jobs, but what makes a job "good"? Pay and benefits are only a part of it. By providing opportunities to pursue interesting work in a supportive environment that promotes creativity, engineering and aerospace fields would be more attractive.[204]

Inspiring

EXPERIENCES IN CHILDHOOD ARE VITALLY IMPORTANT and can provide inspiration for a lifetime, as the biographical information in the first section of this book indicates. The Wright brothers, Robert Goddard, Konstantin Tsiolkovsky, Hermann Oberth, Wernher von Braun, Thomas Edison, and many others each credit encountering inspirational events in their youth with setting their life's work. Their education, also, was often unusual, and they were allowed and encouraged to learn in their own ways.

In 2005, Congress requested the National Academies to investigate what is needed to enhance science and technology in the U.S. The resulting report expressed concern about "a recurring pattern of abundant short-term thinking and insufficient long-term investment" and had several recommendations including: "vastly improv[e] K-12 mathematics and science education;" increase long-term basic research; and "develop, recruit, and retain top students, scientists, and engineers." It was also pointed out that among the nations of the world "the winners ... will be those who develop talent, techniques and tools so advanced that there is no competition." Education must provide the tools—knowledge, understanding, experience—and must provide students with the problem-solving skills to find solutions to challenges.[205]

The pre-teen and teen years are especially critical, when experiences can have life-long effects. A formative experience for many students in 1986 was the *Challenger* accident and in 2003 the *Columbia* accident. These left strong impressions which could discourage those students. However, our response to those setbacks can be more important than the events themselves in teaching young people how to respond to such tragedies. When the Wright brothers learned of Otto Lilienthal's death, they could have concluded that, even if possible, flying was just too risky. To our benefit, though, they became determined to show that flying could be done safely. Their efforts to educate themselves—to understand the key issues, to develop the technical understanding, and to correct the errors of the past—and the support they received to do so, resulted in success.

THE WRITINGS OF JULES VERNE, H. G. WELLS, and other authors of the 19th century provided inspiration to the early rocket scientists. In the 20th century, the writings of Arthur C. Clarke, Isaac Asimov, Edgar Rice Burroughs, and others; radio and television series such as *Buck Rogers in the 25th Century*, *Fireball XL-5*, *Star Trek*, and *The Jetsons*, even; and films such as *2001: A Space Odyssey*, *Space Camp: The Movie*, and the *Star Trek* movies, have served to inspire. Efforts to promote such inspiration should be directed at all students, since we don't know from whom key insights will come to successfully address a challenge.

Inspiring students is also a task of teachers, although, involvement of the parents and community can help greatly. In 1994 in Long Beach, CA, Erin Gruwell, a new English teacher at Woodrow Wilson High School, was assigned to teach "unteachable, at risk" teenagers. She was appalled by the conditions she found there, but through her determination and some transformative moments, the students were given a second chance, and took full advantage of it. Gruwell encouraged her students "to rethink rigid beliefs about themselves and others, to reconsider daily decisions, and to rechart their futures. With [her] steadfast support, her students shattered stereotypes to become critical thinkers, aspiring college students, and citizens for change." With her students, she formed the Freedom Writers Foundation "to equip teachers with the innovative teaching tools they need to reach and empower their students," inspiring others to strive for similar transformation. At its core, transformation is the essence of education, and it is a deeply personal experience.[206]

Education at the higher levels, too, is extremely important, and not just in math, science, and engineering. Ensuring that colleges and universities have the resources they need to provide a quality education for their students across a range of subjects will produce the well-rounded graduates capable of understanding the challenges we face and finding creative solutions. Providing educational opportunities to as broad a swath of the population as possible will also benefit society as a whole. Making knowledge available, in general, is as important as providing a formal education and educational institutions can help with this, too. As a public service, the Massachusetts Institute of Technology provides all of its courses online to anyone, at no charge, through its OpenCourseWare consortium.[207]

Regarding engineering education, according to Domenico Grasso, Melody Burkins, Joseph Helble, and David Martinelli, in an article in *PE The Magazine for Professional Engineers*, the arts and humanities are also important in developing the understanding of how "the physics" relates to the needs of society. Recognition of the importance of having such a well-rounded education is inherent in the "holistic engineering" concept they propose. They say, "the focus today must be on teaching our engineering students to think creatively across scientific, technological, and liberal arts disciplines." Holistic engineering is a cross-disciplinary, whole-systems approach and aims to prepare engineering graduates who understand broader societal issues in addition to being technically proficient, and who can be creative leaders who communicate well. "[T]oo many students perceive traditional engineering degree programs as too prescriptive, lacking breadth and societal engagement, and without connection to pressing issues of social responsibility, entrepreneurial thinking, and environmental awareness—all issues that connect with the youth's hearts and minds." "[T]o attract and retain the best, brightest, most diverse, and most innovative students in the U.S., we must invest in, and actively offer, the highest-quality engineering education filled with integrative courses in engineering technology, humanities, and the arts." They conclude that although "[t]echnologically based engineering training can be outsourced; engineering creativity and innovation, married to technological excellence, cannot. The future of the profession relies upon a core investment in holistic approaches to engineering education, creating the truly 21st century engineering professional who can best meet the complex social, environmental, energy, economic, and technical challenges begging for engineering expertise." "[I]f we train our students to be proficient in engineering thought, as ... holistic thinkers with fundamentals strongly in place as well as the skills to reason, learn, and innovate beyond traditional disciplines, we will have created truly competitive and value-added engineer[s]." "The holistic engineer is, therefore, the most competitive employee of all."[208]

To further develop "the future leaders of the emerging global space community," the International Space University (ISU) was founded in 1987 to provide graduate-level training. ISU was "founded on the vision of a peaceful, prosperous and boundless future through the study, exploration and development of space for the benefit of all humanity." The theme of the 12th Annual ISU Symposium in February 2008 was "Space Solutions to Earth's

Global Challenges," posing the question "How can space address Earth's global challenges in the 21st century?" The "objective was to attract experts from the human spaceflight domain as well as from the Earth and space science communities which, though sometimes seen as being at cross purposes, may well offer different yet complementary solutions to global challenges that we all face." As Charles Cockell, an ISU alum and author, puts it, "environmentalism and space exploration have one and the same objective: to ensure humanity has a home."[209]

Discovering

THE TECHNICAL ADVANCES NEEDED TO ACHIEVE the proposals in Part III do not require violating any of the known laws of physics. They do involve extending our knowledge and capabilities and, so, require focused research and development to produce incremental improvements in existing technologies while looking for breakthrough discoveries.

Researchers at educational institutions, though they may receive funds from corporate sponsors, usually don't have the urgency to produce commercial products that researchers in product development labs face. Academic researchers tend to make more fundamental discoveries like new types of materials, such as aerogels and carbon nanotubes. The task of commercializing the results of fundamental research is typically left to private enterprise. Products made from aerogels and carbon nanotubes, in particular, potentially have enabling roles in space habitats and space elevators, for example, though they could also improve the energy efficiency of buildings (aerogels) or reduce losses in electricity transmission (carbon nanotubes).

Aerogels are the lowest density solids known (sometimes called "solid smoke") yet have a high strength. They are also excellent insulators and were used for insulation on the Mars rover spacecraft. Aerogels were first developed in 1931 by Sam Kistler, a university researcher who later went to work for Monsanto where he continued his work. In labs such as the Aerogel Research Laboratory at the University of Virginia, where Pamela Norris conducts research, discoveries are continuing to advance our understanding of aerogels and their production and uses. In addition to being thermal insulators, aerogels are also electrical and acoustic insulators, and are used for such diverse applications as DNA purification and sleeping bag insulation.[210]

The discovery of carbon nanotubes, described in Chapter 9, is another example of basic research being performed in academic research settings. As greater understanding of the material is gained, products utilizing its' unique properties are becoming available.[lxviii] Yet another example is the recent discovery of "a new chemical pathway that turns carbon dioxide (CO_2) into plastics such as polyurethane or polycarbonate, the most widely used plastics in the world," and that are typically derived from oil. In 2008, Toshiyasu Sakakura, a chemist with the Japanese National Institute of Advanced Industrial Science and Technology in Tokyo, reported developing a catalyst that greatly simplifies the chemical process by combining several steps. Additional work is needed to make the process commercially viable, but the ability to easily make a widely used plastic from a waste product (CO_2) rather than a diminishing resource (petroleum) would contribute to long-term sustainability. Using a different process, Geoffrey Coates of Cornell University in Ithaca, NY, "has made polymers ... that are 30 to 50 percent CO_2." If a process could be developed whereby atmospheric CO_2 could be transformed into such plastic, or into carbon nanotubes, then an abundant undesirable waste product could be made into another highly desirable material.[211]

Such advances are vital to providing the technical options needed to address our challenges. These discoveries must then be developed into products that provide solutions.

[lxviii] "The Project on Emerging Nanotechnologies," Nanotechnology Consumer Products Inventory, The Woodrow Wilson Institute and the Pew Charitable Trusts, http://www.nanotechproject.org/inventories/consumer/

22

Private Enterprise:
Developing Products and Providing Services

"It is not the strongest of the species that survives, nor the most intelligent, but the one most responsive to change." — Charles Darwin

ONE ADVANTAGE ATTRIBUTED TO PRIVATE ENTERPRISE is the ability to adapt quickly to changing circumstances. When considering the whole of enterprises, small and large, this is certainly the case, although businesses that are slower to adapt, or are even resistant to change, will eventually fail. While some companies perform their own basic research to come up with new ideas, more typically their research involves developing products by applying the results of basic research that was either funded or performed by others. The role of developing those products, though, is vital in order to have the best array of solutions to choose from to address a challenge, and private enterprise is an effective means of developing ideas into products. Even companies that do not develop products, however, have an important role by providing services that may use the products developed by others. Through these roles, companies help to create the future.

• • •

IN DECEMBER 1984, DICK RUTAN AND JEANNA YEAGER flew an experimental airplane, *Voyager*, nonstop around the world, without refueling—a world record feat. *Voyager* was a fiberglass/carbon-fiber/Kevlar composite—extremely strong and lightweight—and it was the first airplane to be constructed entirely of such composite materials. It was a twin-engine propeller-driven aircraft that had been designed by Dick's brother, Burt, and the total distance traveled was 42,432 km (26,366 miles). This was a remarkable advance from the Wright brothers' 37 m (120 ft) first flight, eighty-one years earlier.[212]

Burt Rutan was not finished setting world records. On June 21, 2004, his *SpaceShipOne* achieved the feat of taking a person into space when Mike Melvill piloted it to an altitude of 100 km (62 miles) over the Mojave Desert in California. It was "only" a suborbital flight, basically duplicating Alan Shepard's flight of 43 years earlier in *Freedom 7*, but what made this a first is that it was privately financed, not a government project. With the successful second and third flights of *SpaceShipOne* in September and October 2004, piloted by Melvill and Brian Binnie, respectively, Rutan won the $10 million Ansari X Prize. This feat hints at the possibility of tourist flights into space, similar to the early aviation barnstormers selling rides to awestruck passengers.[213]

That is exactly what Sir Richard Branson, founder of Virgin Atlantic airline, wants to do. In 2004, he started a new company, Virgin Galactic, specifically to sell suborbital flights into space. To do that, Rutan is currently building several *SpaceShipTwo* vehicles for Branson that will carry six passengers and two crew at a time. The first *SpaceShipTwo* is expected to be ready for flight testing in 2010, with the first passenger flight in 2011. About 200 people have already paid $200,000 apiece, and about 85,000 have paid registration fees for future flights. Rutan expects that by 2020 over 100,000 people could fly aboard the vehicles, but $200,000 is still out of reach for most people. However, there were 25 other teams from around the world entered in the Ansari X Prize competition, working to lower the cost even more.[214]

Private citizens have already visited space, beginning with Dennis Tito in 2001, when he spent a week on board the *ISS*, launched in a Russian *Soyuz* spacecraft. He was followed by Mark Shuttleworth in 2002, Gregory Olsen in 2005, Anousheh Ansari in 2006, Charles Simonyi in 2007, and Richard Garriott in 2008. (Ansari, who provided the prize money for the Ansari X-prize, has described her space travel experience in her book, *My Dream of Stars*.) Each of them paid at least $20 million for their personal expedition—an amount that is prohibitively expensive for most people.[215]

Rutan's ultimate goal is to enable thousands of people to go to space, including to the Moon, and he is planning *SpaceShipThree* to reach Earth orbit. The suborbital flights of *SpaceShipOne*, though, required only about two percent of the energy needed to get to LEO, and *SpaceShipOne* was not subjected to the intense heating that a reentering orbital vehicle experiences. So, achieving orbital flight will be a major step for private space enterprise. The difficulty of that challenge is reflected in the number of failed . attempts, from the Conestoga Rocket to the Rotary Rocket. But private space enthusiasts are not deterred by the failures—those are simply learning experiences—and at the beginning of the space age in the 1950s and 1960s, many government-funded rockets failed to reach space, too.[216]

Robert Bigelow, owner of the Budget Suites hotel chain and founder of Bigelow Aerospace, is taking a somewhat different approach with plans for a space hotel as a tourist destination, in addition to a research and manufacturing space station. Utilizing technology developed by NASA for inflatable space habitat modules, Bigelow has already launched *Genesis I* and *Genesis II*, sub-scale modules that are currently orbiting Earth, using Russian *Dnepr* rockets. His full-scale *Sundancer* module is expected to be launched in 2010. Currently, however, there are no suitable spacecraft to launch people to these modules. Even the Russian *Soyuz* is too expensive, so Bigelow offered a $50 million America's Space Prize to develop, by January 10, 2010, a spacecraft capable of taking five people to orbit and docking with his space stations. Though several teams attempted to win this prize, none were able to do so by the expiration date.[217]

Another entrepreneur, Elon Musk, was one of the contenders and founded SpaceX to develop just such a vehicle.[lxix] On September 28, 2008 his *Falcon 1* rocket was successfully launched to orbit, after three failed attempts to reach space. *Falcon 1* is designed to carry up to 569 kg (1,254 lb) to LEO for $7.9 million.

[lxix] Musk also founded Paypal, the on-line payment processing service.

That vehicle is not capable of launching people into orbit, but a larger version, *Falcon 9*, will be and Musk wants NASA to use this vehicle for launching astronauts to the *ISS* after the Space Shuttle stops flying in 2010.[218]

Looking even further ahead, Bigelow has also proposed to NASA the idea of assembling his modules at the L1 location between the Earth and the Moon and then delivering them to the Moon as complete Lunar bases, and Bill Stone, founder of Stone Aerospace, intends to process water and fuels on the Moon, transfer them to a LEO refueling station, and provide refueling services to spacefarers on a first-come, first-served basis. Stone anticipates that it will cost $15 billion and, in 2007, formed Shackleton Energy Company to lead the business development effort. He expects to be open for business around 2015.[219]

Through such efforts, entrepreneurs are advancing our capability to economically reach space. As the cost comes down further, the number of people who can get a spacecraft-eye view of the Earth and space will grow from dozens to hundreds to thousands to ???

THOMAS EDISON DID MORE THAN PERFECT the incandescent light bulb and develop the generators and distribution system to light the world. In Menlo Park, NJ, he invented the product development laboratory, his "invention factory." His systematic approach to invention set a new standard and his lab developed the phonograph, motion picture camera, fluoroscope (for medical X-rays), carbon microphone, and a host of other devices that became commercial products. For his efforts he earned the nickname "the wizard of Menlo Park." Today, product development laboratories are a standard part of major corporations and many smaller companies.[220]

Companies, large and small, play important roles in inventing the future. Developing new products or capabilities that provide better ways to do things is one role. Providing services that may not otherwise be available is also an important role. Promoting education is another role, through such endeavors as the Northrup Grumman Foundation Earthwatch Educator program, an environmental education program for teachers. Even companies that are not pushing the edge of science or technology can play a role. Mattel, Inc., for example, makes toys and games that stimulate the imaginations of children, from educational preschool toys to games to dolls, including Barbie®. The original 1959 Barbie doll was a fashion model, with clothing and related accessories. In 2008, Mattel introduced Space Camp Barbie™, part of their "I Can Be..." series promoting interest in math and science. Space Camp Barbie comes with a spacesuit and helmet so girls can imagine being astronauts. Mattel also makes Jimmy Neutron, Boy Genius™ toys, aimed at boys. (But where is Sally Charm,[lxx] Girl Genius?) These efforts help children aspire to technical fields, but there is much more that could be done.[221]

Companies can play more direct roles in creating the future. Google, for example, is investing millions of dollars in renewable energy with a goal of making renewable electricity cheaper than power from coal-fired plants, aiming for 3 cents/kWhr. Initially focusing on buying power from companies producing renewable energy from solar thermal, wind, geothermal, and other technologies, Google has committed $20 million for research and development. This future-oriented vision is also expressed by the launch, in 2007, of the Google Lunar X Prize Competition with $30 million in prizes. The goal is to have, by the end of 2012, a Lunar Rover that roams for at least 500 meters and sends video, images, and data back to Earth. This competition could lead to a spacecraft that would satisfy Proposal 2 in Chapter 20 for a $75 million rover/sample return mission that can go to any location in the inner solar system.[222]

The X Prize Foundation also offers other prizes, with the goal of stimulating "innovation that makes a lasting impact" from technological breakthroughs or using "existing technologies, knowledge or systems in more effective ways." The prizes are intended to "engage multidisciplinary innovators which would otherwise be unlikely to tackle the problems that the prize is designed to address." The prizes are not just space-related. On

[lxx] A Charm is a sub-atomic particle that is a type of quark.

March 20, 2008 at the New York International Auto Show, the Progressive Insurance Company and the X Prize Foundation announced the Automotive X Prize, a $10 million prize for developing a production-capable vehicle that exceeds 100 mpg equivalent and produces less than 200 gram/mi greenhouse gas emissions. Two long-distance races will be held in 2010. The prize value is the same as the Ansari X Prize, indicating that the sponsors feel that improving automobile efficiency is as important and challenging as suborbital space flight.[223]

THE "WHAT'S YOUR CRAZY GREEN IDEA?" CONTEST was initiated by the X Prize Foundation on September 10, 2008, soliciting ideas for an Energy and Environment X Prize. Peter Diamandis, the X Prize Foundation founder, reminded potential entrants that "the day before something is truly a breakthrough, it's a crazy idea." Anyone with an idea was requested to submit it as a 2-minute YouTube video, for a chance to win $25,000. Other organizations also offer prizes for innovative ideas. The Pete Conrad Spirit of Innovation Award encourages high school teams to develop solutions for problems in three areas: Lunar exploration, personal spaceflight, and renewable energy. First place prizes for each area are $10,000. And Richard Branson, in 2006, "declared that all future proceeds from the Virgin Group's transportation companies will be invested into renewable energy initiatives" and announced "a $25 million prize to encourage a viable technology which will result in the net removal of anthropogenic, atmospheric greenhouse gases." The Heinlein Prize, announced on January 19, 2009 and named in honor of Robert Heinlein, the science fiction author, offers a $25,000 grant for a microgravity research competition and the flight of the winning experiment. Such efforts are important for stimulating the creativity of individuals. [224]

23

Your Role: Creativity and Action

"Space is for everybody. It's not just for a few people in science or math, or for a select group of astronauts. That's our new frontier out there, and it's everybody's business to know about space." — Christa McAuliffe, December 6, 1985, teacher-in-space astronaut

"Every single one of us can do something to make a difference." — Archbishop Desmond Tutu

"Never doubt that a small group of thoughtful, committed people can change the world." — Margaret Mead

YOU HAVE A VERY IMPORTANT ROLE in shaping your future and addressing the challenges of the 21st century. In a word, your role is to participate—to be involved, whatever your circumstances, wherever you are in your life. What you can do depends upon your inclination and abilities, but never doubt that you have an effect. As individuals it may seem that we have no significant influence on the world, yet individuals do have influence and, when acting together as a group, that influence is magnified. Organizations depend upon individuals, because the ideas that they champion all come from individuals. So, being creative and acting upon that creativity—which is the way that ideas are developed into solutions—are the most important roles of individuals.

• • •

IN *THE COURAGE TO CREATE*, ROLLO MAY defines creativity as "the process of *bringing something new into being*" (emphasis in the original). Creativity is the act of "enlarg[ing] human consciousness" and "is the most basic manifestation of a man or woman fulfilling his or her own being in the world." Creativity is "in the work of the scientist as well as in that of the artist, in the thinker as well as in the aesthetician; and ... in captains of modern technology as well as in a mother's normal relationship with her child." May then expands the definition by saying that "creativity ... is the encounter of the intensively conscious human being with his or her world." Creativity requires participation. At the end of his book, May says "[t]he creative process ... [is] the struggle to bring into existence new kinds of being that give harmony and integration." [225]

Every idea or insight that ever occurred began with an individual, but those ideas and insights can profoundly change our views of the world. From the first person who tamed fire to cook food and provide heat; to the first Chinese who recognized that the "magic stones" (magnetized lodestones) could be used to determine direction, enabling more accurate exploration of the Earth; to Eratosthenes who measured the Earth; to the first intrepid sailor who sailed away from land; and many, many others, individuals think and invent and act, developing new ideas, accomplishing new feats, creating new ways of doing things. Important insights and advances may come from anyone.

PAINTINGS FROM THE MIDDLE AGES OR EARLIER appear rather flat, often with proportions that do not look quite right to our eyes. Around the year 1415, Filippo Brunelleschi looked at the Baptistry of the Cathedral of Florence, Italy (the Basilica di Santa Maria del Fiore) and saw it in a way that thousands of other people who had passed the same spot had missed. He drew a picture of the Baptistry that was the first known use of linear perspective. His insight was that by using single-point perspective, where parallel lines converge on a single vanishing point, a more realistic rendering could be drawn, providing depth to a two-dimensional image. That insight changed the way that scenes were portrayed by painters, giving them a more lifelike appearance, and Renaissance art vibrantly expresses this insight.[226]

Brunelleschi did not stop there, though. The cathedral needed a dome over the nave and, again, he had an insight that others had missed. The area to be covered was larger than any previous dome, so this would also need to be the tallest dome, and the standard construction technique, using wooden supports from the ground, would not work. Other architects had not solved the dilemma, but Brunelleschi saw a way to do it—by having the dome be self-supporting as it was raised. Completed in 1436, the resulting dome weighs 37,000 tons and contains over 4 million bricks. With a diameter of 42 m (138 ft), it was the largest dome at the time of construction and, almost 600 years later, it is still the largest masonry dome ever built. If you are ever in Florence it is well worth the nominal fee to climb the passageway between the inner and outer shells of the dome for a close-up look of the dome construction. It is an incredible structure.[227]

If you have ever wondered what one individual can do, think about any of the people mentioned in this book. They first learned as much as they could from others, and then extended their knowledge with new insights, and changed the world. As Isaac Newton said, "If I have seen further, it is by standing on the shoulders of giants." These are the individuals who see possibilities, the creative ideas that evade others. Such creative insights will be needed to address the challenges of the 21st century, on Earth and in space. Even if you do not discover some insight with world-changing results, just following your dreams and doing your best can make a difference.

EILEEN COLLINS GREW UP IN A PUBLIC HOUSING project in Elmira, NY. When she was young, she was fascinated by flying and wanted to be a pilot. That might seem to be a dream too far for someone in her situation, but she was determined and did learn to fly, paying for lessons with a pizza parlor job. She attended college, studied mathematics and economics, and then enlisted in the Air Force to fly jets. She rose to the rank of Colonel and, when the opportunity appeared to become a Space Shuttle pilot, she applied and was accepted. She piloted two Space Shuttle missions before becoming the first woman to command a Space Shuttle, *Columbia* on mission STS-93 in July 1999 to deploy the *Chandra X-Ray Observatory*.[228]

John Herrington, a native of Oklahoma and a member of the Chickasaw Nation, became a Navy pilot and then a NASA astronaut. In November 2002, he became the first Native American in space when he was the flight engineer on Space Shuttle *Endeavour* mission STS-113. He performed three spacewalks to assemble the truss on the *ISS*. On August 13, 2008, he began a three month "Rocketrek," riding his bicycle from Washington to Florida, stopping along the way in each state to talk about his experiences and to inspire students "to realize their potential that lies within."[229]

Peggy Whitson grew up on a farm in Iowa, but after receiving degrees in biochemistry became a researcher at the Johnson Space Center. She was also a university professor, but wanted to be an astronaut and she had an opportunity in 1996 when she was selected to be an astronaut candidate. In 2002, she was part of the *Expedition 5* mission to the *ISS*, living in space for 6 months. On October 12, 2007, during *Expedition 16*, her second mission, she became the first woman commander of the *ISS*.[230]

When Homer Hickam was a boy, he and some friends built a rocket that achieved supersonic exhaust velocity, with the help of a machinist in the shop of a West Virginia coal mine where his father was the foreman. He later went on to work for NASA, supporting space missions and training astronauts. His story is well-known for the book he wrote (*Rocket Boys*) and the movie about it (*October Sky*).[231]

Walter Cronkite inspired millions in the 1960s with the excitement he conveyed when reporting on the space program on the evening news. At age 91, his excitement had not abated and on January 31, 2008, the 50th anniversary of the launch of *Explorer I*, Cronkite was scheduled to speak at a special commemoration at the U.S. Space & Rocket Center in Huntsville, AL. It was only because of poor health that he was unable to attend.[232]

WHILE IDEAS AND VISIONS ARE INITIATED by individuals, realizing those ideas and visions usually involves groups. Joining organizations that share your vision can help to realize it, especially through the political influence that organizations can have. Contacting elected officials to let them know your concerns and what legislation or policies you favor is another way to have your ideas be heard, and anyone can write a letter to the editor of newspapers and magazines, or set up a web page, to present his or her views.

You have a vital role in shaping the future. The challenges before us to address energy and environmental issues, and to become a space-faring civilization are considerable, but you may have an answer that can solve them. Knowledge and education are key. Most of all, stay informed, participate, become involved. Our future is what we—collectively and as individuals—make it.

Conclusion

A New Beginning

"If you want to build a ship, don't drum up people to collect wood and don't assign them tasks and work, but rather teach them to long for the endless immensity of the sea." — Antoine de Saint-Exupery

• • •

ABOUT SIXTY THOUSAND YEARS AGO, a group of people crossed the Red Sea from Africa into Asia, beginning a migration that has resulted in people living on every continent—from the hot, humid tropics to the cold, dry polar regions and in orbit around the Earth. As described in a July 2008 article in *Scientific American*, "The reason they left their homeland in eastern Africa is not completely understood. Perhaps the climate changed, or once abundant shellfish stocks vanished." Whatever the specific reason, the risks of staying became greater than the risks of exploring, so they undertook the arduous task of leaving familiar, though no longer acceptable, territory to search for a more agreeable place to call home. They sought a new beginning, heading into the unknown, into the future.[233]

The urge to explore and settle new territories is often borne of necessity, though there are those restless souls among us who are eager to go even without a clear need. Visionaries see the possibilities, explorers (scientists, engineers, and entrepreneurs, as well as adventurers) develop the means and test the limits of those possibilities, and then, with sufficient motivation, settlers make the possibilities into reality—seeking better, more satisfying lives for themselves and their children. Perhaps in the course of migrating they come to realize that "over the horizon" there are new opportunities, so they learn that trying new things is worthwhile and can lead to a better life.

The ultimate expression of this attitude may be what has been defined as the American Dream: "a predisposition to think that current conditions are not enduring, that times will get better, if only we seek to make them so." But this, really, is the Human Dream. Through the ages, in countries around the world, people have endeavored to improve their lives and where they are not constrained by prejudice or restrictive ideology or social segregation, they have been successful. In recent years, this is evident in countries such as China and India and Brazil where new opportunities, not previously available to them, have enabled hundreds of millions of people to raise their standard of living. That is only a small taste of what is possible.[234]

We are at the beginning of new opportunities for humanity.

24

A New Age of Discovery

"The journey of a thousand miles must begin with a single step."
— Lao Tzu

WE ARE ON THE CUSP OF A 21ST CENTURY AGE OF DISCOVERY—
about the Earth, about the solar system, about ourselves, and our
place in the cosmos. In the process of ensuring a sustainable world
with abundant energy and resources we can develop the means to
become a space-faring civilization. The early exploration is
underway, the first pioneers are living in space, and the
technologies needed for greater expansion into space are being
developed. There are abundant opportunities in space, but the only
way to learn of them—and how they may help us to address the
challenges we face on Earth—is to go there.

• • •

IN THE APPALACHIAN FOOTHILLS of northeast Alabama, nestled
among the trees on a hillside overlooking a tributary of the Little
River, Camp Riverview provides a pleasant retreat far from the
noise and haze of cities. On the first weekend of October in 2001,
less than a month after the terrorist attack on September 11, I
attended a dance event at Camp Riverview. The name of the event
was Dance Vortex—"One HOT weekend of Cajun, Zydeco,
Swing, & Waltz Music & Dance"—and the theme was "2001: A
Dance Odyssey" referring to the 1969 movie *2001: A Space
Odyssey*. It was a more appropriate reference than the organizer
planned. At dusk on the evening of October 5, as some of us were
heading to the dance hall across a large open field, we looked up as
the last rays of sunlight were fading from the crystal clear sky and,
against the familiar background of stars beginning to appear, saw a
bright star-like light progressing steadily from west to east. It was

not a high-flying airplane, as sometimes is the case. The light was well above the atmosphere—it was sunlight reflecting from the *International Space Station*, and we watched raptly as it passed, 352 km (220 mi) above us, excited to be seeing it. On board the *ISS*, the *Expedition Three* crew—Frank Culbertson, Vladimir Dezhurov, and Mikhail Tyurin—was preparing for the first of three spacewalks to outfit the new *Pirs* docking compartment and to attach scientific experiments to the outside of the *Zvezda* service module. They were the early pioneers learning to live in the new territory of space at the beginning of a new age of discovery.

THE SPANISH CONQUISTADOR HERNANDO DE SOTO came upon the Little River in the early 16th century, during his trek across what is now the southeast U.S. Not far from Camp Riverview, at a falls named for de Soto, the Little River plunges 32 m (104 ft) into a 198 m (650 ft) deep canyon, the deepest one east of the Mississippi River. To de Soto, the canyon was a barrier in his path. He was not impressed with the natural richness of the land or the flora and fauna. Nor was the cultural richness of the native peoples of interest to him. In fact, much of that cultural richness was lost, either intentionally or inadvertently, due to the actions of de Soto and his fellow conquistadors. De Soto was on a futile and fatal quest for gold.[lxxi]

The Age of Discovery of which de Soto was a part was not just about acquiring material riches or claiming territory. New knowledge and greater understanding of the world was also being acquired. An expedition led by Ferdinand Magellan sailed around the world for the first time, returning to Spain in 1522 with exotic items and tales of adventure, though without Magellan or most of his crew who had died along the way. Over the following centuries discoveries continued with some explorers most interested in scientific knowledge. In the 1830s, naturalist Charles Darwin sailed around the world—loosely following the path of Magellan—and observed the variations of species in far-flung places. His

---

[lxxi] This was de Soto's second expedition, in 1539 and 1540. His first, the conquest of the Incas in the early 1530s, was very profitable for him.

discoveries eventually led him to propose an explanation for the diversity of life.[lxxii] This spirit of discovery continues today.

NASA produces posters to inspire and to document its accomplishments. After the Space Shuttle became operational, there was a series of posters captioned "Going to Work in Space," with photographs of the Space Shuttle during launch or landing or in orbit carrying out its missions, emphasizing the goals of routine operations and commercial activities. Another of my favorites shows advances in transportation with a horse-drawn stage coach and an early automobile, a sailing ship and a paddle-wheel steamboat, railroad trains and a hot air balloon, the Wright Flyer and Lindbergh's *Spirit of St. Louis*, *Skylab* and the *International Space Station*, and rockets from the *Redstone* that launched Gus Grissom to the *Saturn V* moon rocket, all overlaid on the exhaust of a Space Shuttle heading to space. The background shows the Earth, Moon, and Mars, and deep space with the Milky Way splashed across the sky. The caption is "Challenge Traditions," expressing the idea that each advance requires challenging the assumptions of the status quo. Ultimately, advances occur through overcoming the limitations of the present.

Scientific and commercial endeavors are important activities on board the *ISS*, because they may lead to technological advances and improved products. However, the international cooperation that is integral to these efforts to live in a new environment may be even more important. By exploring space we advance all of these aspects. As John Marburger puts it, "Exploration that is not in support of something else strikes me as somehow selfish and unsatisfying, and not consistent with the fact that we are using public funds for this enterprise, no matter how small a fraction of the total budget they may be." One major difference between settling the American West (or, indeed, occupation of many other parts of the Earth throughout history) and settling the solar system is that there are no known indigenous species in space to displace or conquer. The only known natives in the solar system are us, on

---

[lxxii] Darwin detailed his explanation in the book *On the Origin of Species by Means of Natural Selection*, published in 1859.

the Earth. We need only work out acceptable agreements among ourselves. Successfully achieving such agreements could bring about an Age of Discovery far greater than in the 16th century, and one that is based not on conquest, but on cooperation.[235]

Only a handful of technological advances were needed in order to settle the West: the steel plow, food canning, railroads, mining techniques, and the telegraph were major ones. In comparison, settling space will require many more technical advances: radiation protection; life support; microgravity countermeasures; food storage and supply; propulsion systems; communications; computers and robotics; guidance, navigation, and control; waste recycling; and so on. The technical challenges are much greater, so it is not unreasonable to expect that it would take longer to go from the "corps of discovery" stage to "the West is settled" stage. For the West, that was about 85 years. The solar system is not likely to be "settled" (as in no region uninhabited) by 2054, but significant settlement could occur by then, with towns on the Moon, space habitats orbiting the Earth and sun, and outposts on Mars.

By the end of the 21st century, as visionaries have imagined, settlements throughout the solar system could be home to millions of people and efforts to bring Mars to life could be well underway. One hundred years from now engineers could be devising practical applications from string theory or some theory yet to be proposed to explain the nature of the universe. Einstein's goal of a "Unified Field Theory" may yet be attained, connecting electromagnetic forces with the force of gravity, perhaps involving the dark energy of the universe. Controlled fusion reactors may be perfected and the first spacecraft destined for planets orbiting stars light-years away could be well on their way and sending back data and pictures of planets orbiting those stars. Those missions could lead to a new migration of people out of our solar system into the nearby universe, but that is a space program for the 22nd century and beyond. To get there requires dealing effectively with the issues of the 21st century.[236]

25

Crossing the Threshold

"We know what we are, but know not what we may be." —
William Shakespeare

IN A VERY SHORT SPAN OF HUMAN HISTORY, we have come a very
long way, indeed, although for many people the advances are
tantalizingly out of reach or have no direct effect on their lives. We
have also learned that our actions have consequences, some quite
serious. We have the power to destroy all that is precious to us, but
we also have the power to create a more humane, more just, more
sustainable civilization and to take our place in the cosmos while
doing so.

• • •

IN HIS BOOK *COLLAPSE: HOW SOCIETIES CHOOSE to Fail or
Succeed*, Jared Diamond offers cautionary tales on the
consequences of poor decisions, using historical examples of
societies that collapsed. Environmental mismanagement was
significant in the declines of those societies, but we can learn from
them how to avoid their fates. One lesson is the importance of
maintaining the natural environment. An example relating to this
concerns the Rapa Nui society on Easter Island in the Pacific
Ocean, one of the world's most isolated inhabited islands. Known
most famously for their moai stone statues, over time the Rapa Nui
cut down all of the trees of the once heavily forested island, partly
in the process of making and transporting the moai. The resulting
damage due to deforestation led to soil erosion and loss of fertile
land, reducing the ability to grow food. Unlike other places where
similar environmental damage occurred, there were no nearby
societies with which to exchange food or ideas, nor were there
other islands to which to migrate. As a result, the society

degenerated to internecine warfare and collapse. Easter Island is an example of an isolated society that died out due to overconsumption of necessary resources. Diamond compares this with our present situation. On a global scale, we are like the isolated Easter Islanders, with nowhere to go in the event of excessive environmental damage.[237]

One reason to establish settlements in space might be to provide places to go to if the Earth becomes uninhabitable (as in the movie *Wall-E*), but a better reason would be that going into new territory provides a new perspective (as Clarke described in *The Promise of Space*) and new opportunities. With new knowledge, we gain a fuller, more realistic understanding of the universe.

THE *HUBBLE SPACE TELESCOPE* WAS AIMED at a tiny patch of sky in the constellation *Fornax*, near *Orion*, from September 24, 2003 through January 16, 2004. This patch, one-tenth the diameter of the full moon (about the size of the nail of your little finger held at arm's length), appeared to be totally empty. For 1,000,000 seconds (averaging about 2.5 hours each day) over the course of 400 orbits of the Earth, *Hubble* peered at this patch, recording the photons coming from it, though it seemed to be quite dark. The resulting image shows only a handful of distinct, individual stars that are part of our own Milky Way galaxy, but what the *Hubble* Ultra Deep Field view revealed were thousands of entire galaxies. The total count in this one image is over 10,000 galaxies—each made up, on average, of billions of stars—the common spiral and elliptical galaxies, plus "a zoo of oddball galaxies ... like toothpicks" and "links on a bracelet." That image, and others from the *Hubble*, have revealed previously unknown features of the universe, including the first visible-light images of planets orbiting other stars, but there is far more out there that we have not yet discovered. The universe contains wonders unimagined.[238]

The transformations that occur over the 21st century will be at least as dramatic as those witnessed by previous generations. Ensuring sustainability and providing broad opportunities should be the guiding criteria, but we have an opportunity to do more than simply acquire abundant energy and utilize resources without damaging the Earth's ecosystems. Society has transformed considerably over the past 100 years and the challenge before us now is to continue transforming society, in ways that are beneficial for the environment, for individuals, and for society as a whole. This will involve transforming ourselves to be more aware of the consequences of our actions, to be more informed about the possibilities and options available to us, and to be more cooperative in our efforts to address the challenges we face. With these steps, we can have a future in which everyone can live more fulfilling, meaningful, and equitable lives.

Human nature being what it is, achieving energy and resource abundance will not solve all of the world's ills, but it will help by reducing sources of conflict. Nor will a broad program of space activities solve every pressing issue, but, again, cooperative efforts in space will help by encouraging peaceful interaction between nations. Among the greatest challenges will be dealing with those who have different objectives, different values, and who may not be inclined to cooperative action. These difficulties should not dissuade us from making the attempt. We may even be pleasantly surprised at the outcome.

The challenges we face today are solvable, with determined effort, and advancing into space can be an important part of the solution. The Moon, Mars, and other bodies in the solar system may today seem as distant and as daunting as the American West did 200 years ago to those who did not live there. But as our ancestors crossed thresholds to inhabit the Earth we can cross the threshold to become a space-faring civilization, while enjoying the benefits of those efforts. Space is only 100 km (62 miles) away, you just need to look up.

References and End Notes

[1] Padma, T.V., "Climate Change: Serious Security Threat Warns Report," Inter Press Service, December 11, 2007, http://www.ipsnews.net/news. asp?idnews=40428;

"World in Transition - Climate Change as a Security Risk," German Advisory Council on Global Change, Berlin, 2007, http://www.wbgu.de/ wbgu_jg2007_kurz_engl.html

"As South American Rivers Dry Up, Miners Tap Ocean," Reuters News Service, February 22, 2008, http://www.planetark.org/dailynewsstory.cfm/newsid/47097/story. htm

O'Neil, Peter, "Arctic warming could result in armed conflict: naval expert," *The Ottawa Citizen*, February 29, 2008, http://www.canada.com/ottawacitizen/news/ story.html?id=aa8099cd-dee7-4043-8966-4be1b487bcdf

[2] Livingston, David, "Is space exploration worth the cost?," *The Space Review*, January 21, 2008, http://www.thespacereview.com/article/1040/1

[3] M. King Hubbert, http://www.hubbertpeak.com/Hubbert/

David Goodstein, *Out of Gas: The End of the Age of Oil*, W. W. Norton, 2004.

"IPCC Fourth Assessment Report," Intergovernmental Panel on Climate Change, United Nations Environmental Program, 2007, http://www.ipcc.ch/ipccreports/ar4-wg1.htm

Wilson, Edward O., *The Future of Life*, Knopf, 2002.

Carson, Rachel, *Silent Spring*, Houghton Mifflin, 1962. http://www.rachelcarson.org

[4] "Report of the World Commission on Environment and Development," 96th Plenary Meeting, United Nations General Assembly United Nations, A/RES/42/187, 11 December 1987, http://www.un.org/documents/ga/res/42/ares42-187.htm

"27 nations begin switch to energy-efficient bulbs," The Associated Press, September 1, 2009.

Kyoto Protocol, http://unfccc.int/kyoto_protocol/items/2830.php

United Nations Framework Convention on Climate Change, http://unfccc.int/2860.php

McDonough, William, and Michael Braungart, *Cradle to Cradle: Remaking the Way We Make Things*, North Point Press, A division of Farrar, Straus, and Giroux, 2002

[5] Chambers, Matt, "Dubai Crude Output Down Rapidly as Govt Moves To Crimp Decline," Dow Jones Newswires, July 15, 2007. http://www.zawya.com/printstory. cfm?storyid=DN20070715000464&SecIndustries%2FpagOil%20%26%20Gas&l= 060000070715#DN20070715000464

"Google to Become Carbon Neutral by Next Year," Environment News Service, June 20, 2007, http://www.ens-newswire.com/ens/jun2007/2007-06-20-09.asp

[6] The Masdar Initiative http://www.masdaruae.com/

"Masdar Initiative - Worlds First 100% Carbon Free Community," video, http://www.youtube.com/watch?v=ovly1dQGKH4

"Masdar Eco Friendly city of the Future in AbuDhabi UAE," video, http://www.youtube.com/watch?v=m7CuD91BzR0&feature=related

"Work starts on Gulf 'green city'," BBC news, 10 February 2008, http://news.bbc.co.uk/2/hi/science/nature/7237672.stm.

Palca, Joe, "Abu Dhabi Aims to Build First Carbon-Neutral City," NPR, Morning Edition, May 6, 2008, http://www.npr.org/templates/story/story.php?StoryId=90042092

7   Greensburg, Kansas, http://www.greensburgks.org/

"Greensburg: A Story of Community Rebuilding," Planet Green, Discovery channel, 2008/9. http://planetgreen.discovery.com/tv/greensburg/

Finger, Stan, "Lessons from Greensburg," The Wichita Eagle, March 2, 2008. http://www.kansas.com/greensburg/story/328456.html

"Long-Term Community Recovery Plan," Greensburg + Kiowa County, Kansas, August 2007, http://www.greensburgks.org/recovery-planning/long-term-community-recovery-plan/GB_LTCR_PLAN_Final_HiRes.070815.pdf

8   Schultz, Connie, "A compact or big SUV, we all have our reasons," Newhouse News Service, July 7, 2008.

9   Global Energy Network Institute, http://www.geni.org

"There Is No Energy Crisis, There is a Crisis of Ignorance," http://www.youtube.com/watch?v=-fVI3BRBC6o

10   An Inconvenient Truth, http://www.climatecrisis.net/

Stout, David, "Gore Calls for Carbon-Free Electric Power," The New York Times, July 18, 2008, http://www.nytimes.com/2008/07/18/washington/ 18gorecnd.html

"Gore Calls for a Clean Energy 'Revolution' While Congress Warms to Offshore Drilling," Discover Magazine blogs, http://blogs.discovermagazine.com/80beats/ 2008/07/17/gore-calls-for-a-clean-energy-revolution-while-congress-warms-to-offshore-drilling/

Gore, Al, Our Choice: A Plan to Solve the Climate Crisis, Rodale Books, 2009

The Pickens Plan http://www.pickensplan.com/index.php

11   **Plans for Addressing Energy, Resource, and Environmental Issues:**

1. In 2004, the Carbon Mitigation Initiative of Princeton University, directed by Robert Socolow and Stephen W. Pacala, identified 15 strategies that can each reduce global carbon emissions by at least one billion tons per year by 2054. They refer to these strategies as "wedges" and say that implementing only seven wedges will keep the emissions at current levels, and implementing more will reduce CO_2 emissions. The wedges include increasing the efficiency of buildings, cars (to 25 km/L or 60 mpg),

and coal-fired power plants; increasing the capacity of wind and solar energy by 50 and 700 times, respectively; increasing ethanol production by 50 times; capturing and sequestering CO_2 from 800 coal-fired power plants; and doubling the current nuclear power capacity. In addition, to produce hydrogen for the anticipated fuel-cell cars of the future, they propose using 40,000 km^2 (15,000 mi^2) of solar panels or 4 million wind turbines to generate the needed electricity. (http://www.princeton.edu/~cmi/ Winters, Jeffrey, "Wedge Factor," *Mechanical Engineering*, Vol.129, No. 10, pp.31-35.)

2. Also in 2004, Amory Lovins, founder of the Rocky Mountain Institute, and his co-authors, in their book *Winning the Oil Endgame*, emphatically promote energy conservation and efficiency, for which Lovins coined the term "negawatts." One example of how to do this is the HyperCar, made of advanced materials, strong but lightweight, that simplify construction; with an optimized design to reduce inefficiencies such as drag and friction; and powered by a hydrogen fuel cell. They believe that oil use in the U.S. can be eliminated in a few decades and that energy independence is possible by mid-century. They are not proponents of nuclear power, however, mainly because of its cost. They instead propose four integrated steps: double the efficiency of using oil; speed adoption of "superefficient" cars, trucks, and airplanes; provide ¼ of U.S. oil needs from biofuels (cellulosic biofuels); and save half the projected 2025 use of natural gas through increased efficiency.

Ward, Logan, "Amory Lovins: Solving the Energy Crisis (and Bringing Wal-Mart)," *Popular Mechanics*, November 2007, http://www.popularmechanics.com/technology/ industry/4224757.html?series=37

"Masters of Innovation: Amory Lovins," *Business Week*, March 23, 2001. http://www.businessweek.com/bw50/content/mar2001/ bf20010323_307.htm

Lovins, Amory, "The Negawatt Revolution - Solving the CO_2 Problem," Keynote Address at the Green Energy Conference, Montreal, Canada, 1989. http://www.ccnr.org/amory.html

Hawken, Paul, Amory Lovins, and L. Hunter Lovins, *Natural Capitalism: Creating the Next Industrial Revolution*, Back Bay Books, 2000.

Lovins, Amory, "Supersize The Automobile," *Newsweek*, February 25, 2008. http://www.newsweek.com/id/112733

DeJong, Colleen A., "Here Comes Hypercar," *Automotive Design and Production*, undated, http://www.autofieldguide.com/ articles/010106.html

Lovins, Amory B., E. Kyle Datta, Odd-Even Bustnes, Jonathan G. Koomey, and Nathan J. Glasgow, *Winning the Oil Endgame: Innovation for Profits, Jobs, and Security*, The Rocky Mountain Institute, 2004. http://www.oilendgame.com

3. In his 2008 book, *Plan B 3.0: Mobilizing to Save Civilization*, Lester Brown, founder of the Earth Policy Institute, says that improving efficiency is vital and also claims there is enough harnessable wind to satisfy all of our current energy needs plus provide electricity for plug-in hybrid vehicles. By also increasing solar photovoltaic, solar thermal, hydropower, geothermal, and biofuels (from forest industry byproducts,

animal processing waste, crop residues, and urban tree and yard wastes) he sets a goal of reducing carbon emissions by 80% by 2020. He cautions that producing large quantities of bioethanol from corn can raise the price of corn-based food, and that making bioethanol from sugarcane or biodiesel from palm oil often leads to destruction of rainforest. Producing cellulosic ethanol would avoid these concerns, but it may not be able to compete economically for another decade, though cellulosic plant matter can be (and is) burned to power electricity generation.

Earth Policy Institute, http://www.earth-policy.org/

From August 2006 to August 2007, 6,900 square miles of Amazonian rainforest in Brazil alone, were deforested, largely for growing soybeans and other agricultural activities. Barrioneuvo, Alexei, "Brazil Rainforest Analysis Sets Off Political Debate," *The New York Times*, May 25, 2008, http://www.nytimes.com/2008/05/25/world/americas/25amazon.html

Brown, Lester R., *Plan B 4.0: Mobilizing to Save Civilization*, Earth Policy Institute, 2009, http://www.earth-policy.org/index.php?/books/pb4

4. In his 2007 book, *Carbon-Free and Nuclear-Free: A Roadmap for U.S. Energy Policy*, Arjun Makhijani, a former adviser to the Tennessee Valley Authority (TVA) public utility company, presents a plan for providing the energy we need without relying on fossil-fuels or nuclear energy. He says "[i]t is technologically and economically feasible to phase out CO_2 emissions and nuclear power at the same time" by 2050 or sooner. Increased efficiency, conservation, and use of renewable energy, especially biofuels and solar photovoltaic generation, are key features of Makhijani's detailed roadmap through 2030. He concludes that "a zero-CO_2 U.S. economy without nuclear power is not only achievable—it is necessary for environmental protection and security."

Makhijani, Arjun, *Carbon-Free and Nuclear-Free: A Roadmap for U.S. Energy Policy*, Nuclear Policy Research Institute and the Institute for Energy and Environmental Research, IEER Press and RDR Books, 2007. http://www.ieer.org/carbonfree/index.html

5. At a G-8 meeting in 2009, the International Partnership for Energy Efficiency Cooperation (IPEEC) was formed to promote improvements in global energy efficiency while encouraging commercialization of energy efficiency technologies. By sharing information, best practices, policies, and measures that support energy efficiency efforts among the participating countries, the IPEEC supports a market-based approach to reduce global greenhouse gas emissions. "Secretary Chu Joins with World Leaders to Sign International Partnership for Energy Efficiency Cooperation," U.S. Department of Energy Press Release, May 24, 2009, http://www.energy.gov/news2009/7420.htm

"Joint Statement by the G8 Energy Ministers, the European Energy Commissioner, the Energy Ministers of Brazil, China, Egypt, India, Korea, Mexico, Saudi Arabia, and South Africa," G8 Energy Ministers Meeting, Rome, Italy, May 24, 2009, http://www.g8energy2009.it/pdf/ Session_I_+EC.pdf

[12] *World Energy Outlook 2006*, International Energy Agency, http://www.iea.org/Textbase/npsum/WEO2006SUM.pdf

[13] *On the Threshold of Space,* movie review, *The New York Times*, March 31, 1956, http://movies.nytimes.com/movie/review?res=9E02E3D71E3BE23ABC4850DFB566838D649EDE

[14] White, Frank, *The Overview Effect: Space Exploration and Human Evolution*, American Institute of Aeronautics and Astronautics, 1998.

[15] King, Martin Luther, Jr., "The 'I Have a Dream' Speech," August 23, 1963. http://www.usconstitution.net/dream.html, Video of Speech on YouTube, http://www.youtube.com/ watch?v=PbUtL_0vAJk

The Civil Rights Act was passed on July 2, 1964. http://usinfo.state.gov/usa/infousa/laws/majorlaw/civilr19.htm

Friedan, Betty, *The Feminine Mystique*, W. W. Norton & Company, republished 2001. http://www.americanwriters.org/works/feminine.asp

Fox, Margalit, "Betty Friedan, Who Ignited Cause in 'Feminine Mystique,' Dies at 85," *The New York Times*, February 5, 2006.

JFK Assassination Records, The National Archives, http://www.archives.gov/research/jfk/index.html

White, Jack E., "The Time 100: Martin Luther King," *Time*, April 13, 1998, http://www.time.com/time/time100/leaders/profile/king. html

"Robert F. Kennedy: Biography," Spartacus Educational, http://www.spartacus.schoolnet.co.uk/USAkennedyR. htm

"Beatles' 'Ed Sullivan' appearance rated rock's top TV moment," Associated Press, July 25, 2000. http://archives.cnn.com/2000/SHOWBIZ/TV/07/25/tv.rocks.ap/;

Woodstock '69, http://www.woodstock69.com

"The National Environmental Policy Act of 1969, as Amended," http://ceq.hss.doe.gov/Nepa/regs/nepa/nepaeqia.htm

Fuller, Buckminster, *Operating Manual for Spaceship Earth*, 1963, ISBN 0-525-47433-1.

Star Trek, website http://www.startrek.com/startrek/view/index.html

2001: A Space Odyssey was based on the novel of the same title by Arthur C. Clarke. http://www.imdb.com/title/tt0062622/

[16] President Richard M. Nixon requested a lower budget for NASA in 1971 than it received in 1970. "Statement About the Future of the United States Space Program" March 7, 1970. http://www.presidency.ucsb.edu/ws/index. php?pid=2903

The first and second stages from one Saturn V were used for *Skylab*. Saturn V vehicles assembled from flight and engineering versions are now on display at the Kennedy Space Center in Florida (flight and test stages), at the U.S. Space & Rocket Center in

Alabama (engineering test stages), and at the Johnson Space Center in Texas (all flight stages). http://en.wikipedia.org/wiki/Saturn_V

Launched in 1973, *Skylab* hosted three crews of three astronauts with missions lasting up to 84 days. They demonstrated the ability to live in space for much longer periods of time than previous missions, and performed a wide variety of experiments that provided greatly increased understanding of materials properties, solar astronomy, physiological responses to microgravity, Earth observation, and other scientific fields.

Among the robotic missions *Pioneer 10* and *11*, launched in 1972 and 1973, were the first spacecraft to pass through the asteroid belt, on their way to the outer planets and interstellar space. *Voyager 1* and *2* were launched in 1977 on survey missions of Jupiter, Saturn, Uranus, and Neptune.

Heppenheimer, T.A., *The Space Shuttle Decision: NASA's Search for a Reusable Space Vehicle*, NASA SP-4221, National Aeronautics and Space Administration, Washington, D.C., 1999. http://history.nasa.gov/SP-4221/sp4221.htm

[17] "30th Anniversary of Cuyahoga River Fire," *WCPN*, June 22, 1999. http://www.wcpn.org/news/1999/0622cuyahoga-fire.html,

Gordon, John Steele, "The American Environment," *American Heritage*, Vol. 44, Issue 6, October 1993. http://www.americanheritage.com/articles/magazine/ah/1993/6/1993_6_30.shtml

Environmental legislation and the history of the EPA: http://www.epa.gov/air/caa/peg/, http://www.epa.gov/earthday/history.htm, http://www.epa.gov/history/, http://www.epa.gov/region5/water/cwa.htm

Brigham, Robert K., "Battlefield Vietnam: A Brief History," http://www.pbs.org/battlefieldvietnam/history/index.html http://www.historyplace.com/unitedstates/vietnam/index-1969.html

Weissman, Rozanne, and Ronnie Kweller, "As 30th Anniversary of Oil Embargo Approaches Alliance to Save Energy Examines Then, Now; Says OPEC Still in 'Driver's Seat'," Science Blog, September 29, 2003, http://www.scienceblog.com/community/older/archives/K/0/pub0099.html

Stobaugh, Robert, and Daniel Yergin, editors, *Energy Future: Report of the Energy Project at the Harvard Business School*, Random House, New York, 1979

Braiker, Brian, "Crude Awakening," *Newsweek*, February 17, 2004 http://www.msnbc.msn.com/id/4287300

Strategic Arms Limitation Talks between the U.S. and the USSR. http://www.fas.org/nuke/control/salt1/text/salt1.htm

At the Three Mile Island nuclear power plant in Ohio, on March 28, 1979 a partial core meltdown occurred, leading to the evacuation of thousands of people from their homes in the surrounding area. http://www.threemileisland.org/, http://www.nrc.gov/reading-rm/ doc-collections/fact-sheets/3mile-isle.html

[18] The crime prevention program was called U.N.I.C.O.R.N. for United Neighborhoods Incorporated for Criminal Opportunity Reduction Now! and was funded through the Law Enforcement Assistance Administration (LEAA) and the Comprehensive Employment and Training Act (CETA).

The Junior Engineering and Technical Society, http://www.jets.org

[19] University of Louisville, J. B. Speed Engineering School, http://speed.louisville.edu/cms/content.php

[20] Montreal Protocol on Ozone depletion of the atmosphere, http://ozone.unep.org/Ratification_status/evolution_of_ mp.shtml

Chernobyl.info, http://www.chernobyl.info/

"Chernobyl, 22 Years Later: Exploring the Rubble of the World's Largest Nuclear Disaster," CBSNews.com, March 31, 2008, http://www.cbsnews.com/stories/2008/03/31/eveningnews/main3984592.shtml

The Strategic Arms Reduction Treaty (START I) was signed on 31 July 1991 after a decade of negotiations, to reduce the number of nuclear missiles in the USSR and U.S. arsenals by 30 to 40%, to no more than 1,600 ballistic missiles and 6,000 nuclear warheads. http://www.au.af.mil/au/awc/awcgate/acda/starttex.htm, http://www.dod.mil/acq/acic/treaties/start1/index.htm

The Iran-Iraq War, http://www.infoplease.com/ce6/history/A0825449.html

Segal, David, "The Iran-Iraq War: A Military Analysis," *Foreign Affairs*, Summer 1988, http://www.foreignaffairs.com/articles/43387/david-segal/ the-iran-iraq-war-a-military-analysis

"20th Anniversary of the Exxon Valdez Oil Spill," Exxon Valdez Oil Spill Trustee Countil, http://www.evostc.state.ak.us/ As of 2008, the environment has not completely recovered from the damage, though remaining damage is largely below the surface. *All Things Considered*, National Public Radio, January 23, 2008.

"Exxon Valdez Oil Spill 20 Years Later," CBS News, February 2, 2009, http://www.cbsnews.com/stories/2009/02/02/eveningnews/main4769329.shtml

[21] "*Spacelab* joined diverse scientists and disciplines on 28 Shuttle missions," March 15, 1999   http://science.nasa.gov/newhome/headlines/msad15mar99_1.htm

"The *Hubble Space Telescope*," NASA, http://hubble.nasa.gov/

Hubblesite, http://hubblesite.org/

Voyager, http://voyager.jpl.nasa.gov/science/saturn.html

[22] The *Hubble* mirror aberration that was discovered during checkout on orbit was due to a manufacturing flaw, not a design error. The primary mirror had been precisely ground to the wrong shape, reducing the quality of images. Corrective lenses were installed during the first maintenance mission.

Stephens, Sally, "The New and Improved Hubble Space Telescope," Astronomical Society of the Pacific, The Universe in the Classroom, No. 26, Winter 1994, http://www.astrosociety.org/education/publications/tnl/26/ 26.html

[23] Stenger, Richard, "Man on the moon: Kennedy speech ignited the dream," CNN, May 25, 2001. http://archives.cnn.com/2001/TECH/space/05/25/kennedy.moon/

Kennedy, John F., "Special Message to the Congress on Urgent National Needs," May 25, 1961. http://www.presentationhelper.co.uk/kennedy_man_on_the_moon_speech.htm

An estimated 400,000 people were involved with the Apollo program, which cost about $24 billion. http://www.cbsnews.com/stories/2005/11/03/60minutes/main1008288.shtml

Thimmesh, Catherine, *Team Moon: How 400,000 People Landed Apollo 11 on the Moon*, Houghton Mifflin, 2006

[24] Burns, Ken, "Lewis & Clark: The Journey of the Corps of Discovery," Public Broadcast System, http://www.pbs.org/lewisandclark/archive/1806.html

[25] "Space Settlement Nexus," National Space Society, http://www.nss.org/settlement/

"Space Settlements: Spreading life through the solar system," National Space Society, http://www.nss.org/settlement/nasa/ossws.html

Chandler, David, "A new vision for people in space," MIT News Office, December 16, 2008, http://web.mit.edu/newsoffice/2008/mindell-space-1216.html

Pullen, Lee, "Human Spaceflight Should Drive Evolution," Astrobiology magazine, 22 January 2009, http://www.space.com/scienceastronomy/090122-am-human-spaceflight-evolution.html

"The Future of Human Spaceflight," Space, Policy, and Society Research Group, Massachusetts Institute of Technology, December 2008, http://web.mit.edu/mitsps/MITFutureofHumanSpaceflight.pdf

O'Neill, Gerard K., *The High Frontier: Human Colonies in Space*, Apogee Books, Collector's Guide Publishing, Inc., 3 rd edition, 2000 (first edition 1977).

Zubrin, Robert, *Entering Space: Creating a Spacefaring Civilization*, Jeremy P. Tarcher/Putnam, New York, 1999.

Report from the Presidential Space Task Group, 1969. Cited in Compton, David, *Where No Man Has Gone Before*, NASA SP-4214, NASA, 1989.

Tierney, John, "Earthly Worries Supplant Euphoria of Moon Shots," *The New York Times* on the Web, July 20, 1994. http://partners.nytimes.com/library/national/science/0702094sci-nasa-moon-2.html

Ohio, Indiana, and Illinois were considered the "northwest" in the early 1800s. Jefferson thought it would take many generations to settle the Louisiana Territory and that the native peoples would be gradually assimilated into American society in the process. The idea of Manifest Destiny of the United States to extend across the continent developed in the 1840s and gained broad support over the following decades. [http://en.wikipedia.org/wiki/Manifest_Destiny] Settlement of the West was also delayed due to conflicts over the issue of slavery.

[26] "The History of the Upper Midwest: An Overview," Library of Congress, http://memory.loc.gov/ammem/umhtml/umessay5.html

The first known usage of the term "manifest destiny" was by John L. O'Sullivan in an 1845 essay, though the attitude expressed by the term arose decades earlier. O'Sullivan, John L., "A Divine Destiny for America," http://www.newhumanist.com/md4.html, http://www.classcoffee.com/combo/combo_docs/combo_unit4/manifest_destiny_reading.RTF

[27] "Completing the Transcontinental Railroad, 1869: Driving the Golden Spike," Eyewitness to History.com, http://www.eyewitnesstohistory.com/goldenspike.htm

"June 4, 1876 Express train crosses the nation in 83 hours," This Day in History, History.com, http://www.history.com/this-day-in-history.do?action=Article&id= 4541

[28] Drache, Hiram M., "The Impact of John Deere's Plow," Northern Illinois University, http://www.lib.niu.edu/ipo/2001/iht810102.html

John Deere Company website http://www.deere.com/en_US/compinfo /history/index.html

"About Canned Food: Whence it Came: The History of Food Canning," Food Reference.com, http://www.foodreference.com/html/artcanninghistory.html

Bellis, Mary, "The History of Barbed Wire," About.com, http://inventors.about.com/library/inventors/blbarbed_wire.htm

"History of the American Water Pumping Windmill," Ironman Windmill Company, http://www.ironmanwindmill.com/windmill%20history.htm, Windmills for pumping ground water were invented by Daniel Halladay in 1854.

Baker, T. Lindsay, "Brief History of Windmills in the New World," Windmillers' Gazette, http://www.windmillersgazette.com/history.html

Hooton, LeRoy W., Jr. "Welcoming Remarks," WEF/AWWA Joint Management Conference, Water Environment Federation and the American Water Works Association, March 2, 2006, http://www.ci.slc.ut.us/utilities/NewsEvents/news2006/ news332006.htm The first major irrigation of the West was performed by the Mormons in Utah, which receives an annual average of 13 inches of precipitation each year.

[29] Settlers also established cotton plantations in Texas or other territories where slavery was allowed. Efforts to increase the number of slave-holding states were key to expansion into the southwest and an underlying motivation for the Mexican-American War. "The Handbook of Texas Online," http://www.tshaonline.org/handbook/ online/articles/SS/yps1.html

The Mormons were among the earliest group to settle in the West, and founded Salt Lake City in 1847, to escape the religious persecution they experienced in Illinois and other places they lived. "Mormon History," http://www.historyofmormonism.com/mormon-history

[30] "Homestead National Monument: History and Culture," National Park Service, http://www.nps.gov/home/historyculture/index.htm

"The Homestead Act: Creating Prosperity in America," Legends of America, http://www.legendsofamerica.com/AH-Homestead.html

"Teaching with Documents: The Homestead Act of 1862," http://www.archives.gov/education/lessons/homestead-act/

"Archeology of an Exoduster Neighborhood: The History and Development of the Monroe School Neighborhood," National Park Service, http://www.nps.gov/mwac/brvb/history.htm

"African-Americans in the Old West," IMA Hero Reading Program, http://www.imahero.com/readingprogram/afamwest.html,

"Benjamin 'Pap' Singleton," The Tennessee Encyclopedia of History and Culture, http://tennesseeencyclopedia.net/imagegallery.php?EntryID=S041

"Where Did African-Americans Live?," nebraskastudies.org, http://www.nebraskastudies.org/0500/frameset_reset.html?http://www.nebraskastudies.org/0500/stories/0504_0101.html

"Was the Homestead Act Colorblind?," nebraskastudies.org, http://www.nebraskastudies.org/0500/frameset_reset.html?http://www.nebraskastudies.org/0500/stories/0501_0205.html,

The Mining Act extended the provisions of the 1866 Mining Law, that established mining rights. It has been amended several times, but is still in effect in 2010. Extraction of oil, coal, and natural gas usually requires payment of royalties to the federal government, but not hard rock deposits. Updates to the Act to include hard rock royalties, local community rights, and environmental considerations were presented to Congress in legislation such as H.R. 2262, the Hardrock Mining and Reclamation Act of 2007.

"At Last, an Overhaul For a Bad Bill," October 27, 2007, editorial, *The New York Times*, http://www.nytimes.com/2007/10/26/opinion/ 26fri3.html?_r=1&oref=slogin

Kohler, Judith, "Mining law reform called long overdue," The Associated Press, December 10, 2007.

Danowitz, Janwe, and Richard Wiles, "Our grandest assets under siege," McClatchey Tribune News Service, March 3, 2008. Mining approved by Forest Service within a few miles of the Grand Canyon.

[31] Powell, John, "Encyclopedia of North American Immigration," Facts on File Library of American History, Facts on File, Inc., New York, NY, 2005.

[32] Reflecting the rapid rate of change in the American West following the Civil War, the population of Kansas increased from 107,206 in 1860 to 364,399 in 1870 to 996,096 in

1880 to 1,427,096 in 1890. The population of Iowa increased from 192,214 in 1850 to 674,913 to 1,194,020 to 1,624,615 to 1,911,896 for succeeding census decades. Other western states showed similar rapid increases in population.
http://www.census.gov/dmd/www/resapport/states/kansas.pdf,
http://www.census.gov/dmd/www/resapport/states/iowa.pdf

In 1830 to produce 100 bushels (5 acres) of wheat required 250 to 300 labor hours. By 1890 the same amount of wheat required only 40 to 50 labor hours.
http://www.agclassroom.org/gan/timeline/farm_tech.htm
http://xroads.virginia.edu/~drbr/corporat.html

[33] Evans, Harold, *They Made America*, Little, Brown, and Company, 2004.

Lilienthal, Otto, *Birdflight as the Basis for Aviation*, Markowski International Publishers, 2000

[34] Means, James, "The Aeronautical Annual," 1895 to 1897,
http://www.flyingmachines.org/means.html

[35] Vivian, E. Charles, "A History of Aeronautics," Chapter X - Samuel Pierpoint Langley, 1920. http://worldwideschool.com/library/books/tech/engineering/AHistoryofAeronautics/chap10.html

"Efforts at Powered Flight During the Last Decade before the Wright Brothers," U.S. Centennial of Flight Commission,
http://www.centennialofflight.gov/essay/Prehistory/Last_Decade/PH5.htm

[36] "The Unlikely Inventors," Wright Brothers Flying Machine, NOVA,
http://www.pbs.org/wgbh/nova/wright/inventors.html

[37] "The First Reporter: 'Dear friends, I have a wonderful story to tell you...'," Wright Brothers' Flying Machine, NOVA,
http://www.pbs.org/wgbh/nova/wright/reporter.html

Vivian, E. Charles, "A History of Aeronautics," Chapter XI - The Wright Brothers, 1920. http://worldwideschool.com/library/books/tech/engineering/AHistoryofAeronautics/chap11.html

[38] Evans, Harold, *They Made America*, Little, Brown, and Company, 2004.

[39] Hansen, James, *Spaceflight Revolution: NASA Langley Research Center From Sputnik to Apollo*, NASA SP-4308, p. 3, 1995. The NACA enabling act was attached as a rider to a naval appropriations bill.

Bilstein, Roger E., *Orders of Magnitude: A History of the NACA and NASA, 1915-1990*, NASA SP-4406, NASA, 1989. http://www.hq.nasa.gov/office/pao/History/SP-4406/contents.html

On May 15, 1918 the first continuous scheduled public-service airmail in the U.S. began between New York City and Washington, D.C., via Philadelphia.
http://www.airmailpioneers.org/history/Sagahistory.htm

The Kelly Act was passed in 1925, guaranteeing airmail routes to commercial carriers.

40 "Barnstormers," U.S. Centennial of Flight Commission,
http://www.centennialofflight.gov/essay/Explorers_Record_Setters_and_Daredevils/
barnstormers/EX12.htm

41 Lindbergh was the first to fly solo across the Atlantic, but others had crossed years
earlier. On May 16, 1919 LCdr. Albert C. Read and Walter Hinton, pilot, departed
New Foundland in a Navy seaplane, arriving in Lisbon, Portugal on May 27, 1919,
after stopping at the Azores Islands due to poor weather. Their total flying time was 27
hours. A couple of weeks later the British team of John Alcock and Albert Brown were
the first to cross without stopping, winning a $50,000 prize. They landed in Ireland on
June 16, 1919 after a 16.25 hour flight from New Foundland, a distance of 1,900 miles.
(This was a much shorter distance than Lindbergh's 3,600 mile flight that took 33.5
hours.) From July 2 to 13, 1919, a British dirigible (the R-34) flew round-trip between
Great Britain and New York, taking 108 hours for the westbound leg and 75 hours for
the return.

http://history.acusd.edu/gen/WW2Timeline/firstflight.html
http://www.infoplease.com/ipa/A0004537.html

History of the Lindy Hop Dance, http://www.lindycircle.com/history/lindy_hop/

42 Laminar air flow, http://www.aviation-history.com/theory/lam-flow.htm

In 1928 Frank Whittle, while at the RAF College at Cranwell, England, proposed in his
senior thesis, Future Developments in Aircraft Design, use of a reaction propulsion
system. By 1930 he had developed the concept into the turbojet and received a patent.
The engine was successfully bench tested in April 1937 and test flown in an airplane in
1941.

http://www-g.eng.cam.ac.uk/125/achievements/whittle/telgraph.htm,
http://www.asme.org/communities/History/Resources/Whittle_Frank.cfm]

Independent of Whittle, Hans von Ohain in Germany also developed a turbojet engine,
patenting his design in 1934. By August 27, 1939, the engine was flown in the Heinkel
He178, the first jet airplane to fly. During the first test flight the He178 reached a
speed of 650 km/h (403 mph), faster than any other airplane at that time.
[http://inventors.about.com/library/ bljetengine.htm] This led to the first operational jet
fighter, the Messerschmidt Me 262, which first flew on July 18, 1942. The Heinkel He
280 was the first jet fighter plane, when it flew on March 30, 1941, though it was not
put into operation. The Me262 had an operational speed of 805 km/h (500 mph),
almost 240 km/h (100 mph) faster than the fastest Allied airplanes, and in combat
operations Me262s shot down over 150 Allied aircraft. [With the loss of about 100
Me262s. http://en.wikipedia.org/wiki/Messerschmitt_Me_262] The British Gloster
Meteor jet fighter first flew on March 5, 1943 and entered combat on July 27, 1944.
The U.S. P-80 Shooting Star first flew on January 8, 1944, but the war ended before it
saw combat. ww2.guide.com, http://www.ww2guide.com/jetrock.shmtl

43 Levine, Arnold S., Managing NASA in the Apollo Era, NASA SP-4102, p. 10, NASA
History Series, National Aeronautics and Space Administration, 1982.
http://history.nasa.gov/SP-4102/sp4102.htm

44 Government airmail contracts were implemented as a way to encourage the fledgling airplane industry at a low cost to the government. World Wars I and II brought military issues for which airplanes were uniquely suited. Private efforts such as the Orteig prize and the Collier Trophy also spurred advances, and airplane manufacturers such as the Wrights, Curtiss, Ford, Boeing, Douglas, Beech, Cessna, Martin, De Havilland (British), and others improved the comfort and capabilities of airplanes for the public, as well as the military, while incorporating the NACA advances in the performance and efficiency of airplanes.

1926 Air Commerce Act, http://www.centennialofflight.gov/essay/Government_Role/ 1925-1929_airmail/POL5.htm

Through legislative reforms such as the Civil Aeronautics Act of 1938, regulation and aviation policy-making were assigned to a single agency, with a mission "to preserve order in the industry, holding rates to reasonable levels while at the same time nurturing the still financially-shaky airline industry by protecting carriers from unbridled competition."

http://www.geocities.com/CapeCanaveral/4294/history/1935_1950.html

Also the 1958 Federal Aviation Act, http://www.faa.gov/about/history/brief_history/

The Boeing 247, 12-passenger airplane, introduced in 1933, and the Douglas Aircraft, DC-3, 21-passenger air plane, introduced on December 17, 1935, greatly improved passenger and commercial air travel. In 1952 the British de Havilland Comet was the first commercial jet-powered aircraft. However, it only flew until 1954, when two back-to-back crashes occurred due to design problems. [The failures occurred due to metal fatigue and were related to the use of square windows, which have higher stresses than rounded windows. [http://plane-truth.com/comet.htm]] The Boeing 707, with improvements over the Comet, began transatlantic service on October 26, 1958 for Pan American Airways.

In 1936, airline passengers flew 439 million miles. (Holland, Kevin, "The C&O Air Force: How a railway became a corporate aviation pioneer," Chesapeake & Ohio Historical Magazine, April 2001.) By 1958 this had increased to 24 billion passenger miles. (Loening, Grover, "American Planes: The Lessons of History," *The Atlantic Monthly*, June 1959) This corresponds to over 3 billion "Revenue Passenger Ton Miles" as reported to the Department of Transportation by Large Certificated Air Carriers for 1958. This number increased to almost 13 billion in 1968; 24 billion in 1978; 44 billion in 1988; 64 billion in 1998, and over 81 billion in 2006.

(http://www.bts.gov/oai/indicators/top.html, http://www.bts.gov/programs/airline_ information/air_carrier_traffic_statistics/airtraffic/annual/1954-1980.html)

45 Koestler, Arthur, *The Sleepwalkers: A History of Man's Changing Vision of the Universe*, The Universal Library, Grosset & Dunlap, New York, 1959.

46 The date when Goddard had that vision was so important to him that, in his journals, he referred to October 19 as his "Anniversary Day." http://www-istp.gsfc.nasa.gov/ stargaze/Sgoddard.htm, "Dr. Robert H. Goddard: American Rocket Pioneer," NASA, December 1, 2004. http://www.nasa.gov/centers/goddard/about/dr_goddard.html

[47] Goddards work was funded by a $5,000 grant from the Smithsonian Institution in 1917 and his report was published as Smithsonian Miscellaneous Publication No. 2540 in January 1920. http://rocketsciencebooks.home.att.net/goddard-extreme.html

Kluger, Jeffrey, "Robert Goddard," *Time*, March 29, 1999. *The New York Times*, in an editorial on January 13, 1920, criticized Goddard for claiming that rockets could operate in a vacuum, among other criticisms. On the eve of the first Apollo landing on the Moon in 1969 *the New York Times* printed a retraction. http://www.time.com/time/time100/scientist/profile/goddard.html

[48] "Space prophet Konstantin Tsiolkovsky," *Russia Today*, June 5, 2009, http://russiatoday.com/Top_News/2009-06-05/Space_prophet_Konstantin_ Tsiolkovsky.html

[49] Launius, Roger D. *Frontiers of Space Exploration*. Westport, Conn.: Greenwood Press, 1998

"Herman Oberth," U.S. Centennial of Flight Commission, http://www.centennialofflight.gov/essay/SPACEFLIGHT/oberth/SP2.htm

Price, Michael, "Woman in the Moon," Senses of Cinema, March 2004, http://www.sensesofcinema.com/contents/cteq/04/woman_in_the_moon.html

"Wernher von Braun," NASA MSFC History Office, http://history.msfc.nasa.gov/vonbraun/index.html

[50] Many of the rocketry advances had been developed and patented by Goddard and were in the public domain. (Copies of Goddards reports reached European scientists.).

"Space: Huntsville's History and Legacy," Osher Lifelong Learning Institute, University of Alabama in Huntsville lecture series, October 1 to November 12, 2006

Mueller, F. K., "A history of inertial guidance," British Interplanetary Society, Journal (Astronautics History), ISSN 0007-084X, vol. 38, April 1985, p. 180-192.)

[51] "The International Geophysical Year," the National Academies, http://www.nas.edu/history/igy/

Siddiqi, Asif A., "Korolev, Sputnik, and The International Geophysical Year," from *Sputnik, The 50th Anniversary*, 2008, http://history.nasa.gov/sputnik/siddiqi.html

"Project Vanguard: Why It Failed to Live Up to Its Name," *Time*, October 21, 1957. http://www.time.com/time/magazine/article/0,9171,937919,00.html,

Green, Constance McLaughlin, and Milton Lomask, *Vanguard: A History*, NASA SP-4202, National Aeronauticss and Space Administration, Washington, D.C., 1970. http://www.hq.nasa.gov/office/pao/History/SP-4202/cover.htm

Wilford, John Noble, "With Fear and Wonder it Its Wake, Sputnik Lifted Us Into the Future," *The New York Times*, September 25, 2007. http://www.nytimes.com/2007/ 09/25/science/space/25sput.html?_r=1&ref=science&oref=slogin

52 For more on Korolev and *Sputnik*, see:
http://www.russianspaceweb.com/korolev.html,
http://history.nasa.gov/sputnik/siddiqi.html

Harford, James J., *Korolev: How One Man Masterminded the Soviet Drive to Beat America to the Moon*, John Wiley, New York, 1997.

53 In May 1961 a problem occurred when NASA tested a Redstone rocket with a Mercury capsule in preparation for Alan Sheppards flight. It was decided to launch another chimpanzee to ensure that the problem was resolved, before launching Sheppard. During that delay the USSR launched Gargarin. (Ed Buckbee, "The Real Space Cowboys," session 5 of "Space: Huntsville's History and Legacy," University of Alabama in Huntsville, October 29, 2006.)

President John F. Kennedy, "Special Message to the Congress on Urgent National Needs," May 25, 1961. http://www.jfklibrary.org/Historical+Resources/Archives/Reference+Desk/Speeches/JFK/003POF03NationalNeeds05251961.htm

54 In 1969 Werhner von Braun presented a plan using Saturn V technology with nuclear-powered upper stages to send people to Mars by 1982. (Ed Buckbee, "The Real Space Cowboys," session 5 of "Space: Huntsville's History and Legacy," UAH, Huntsville, AL, October 29, 2006)

55 An enormous variety of achievements in space resulted from that race: discovery of the Van Allen radiation belt (1958, *Explorer 1*, US), discovery of the solar wind (1959, *Luna 1*, USSR), the first pictures of the far side of the Moon (1959, *Luna 3*, USSR), the first weather satellite (1960, *Tiros 1*, US), the first Earth-observation satellite (1964, *Nimbus 1*, US), the first communications satellite (1962, *Telstar 1*, US), the first person in space, Yuri Gagarin (1961, *Vostok 1*, USSR), the first extravehicular activity in space (1965, *Voskhod 2*, USSR), the first rendezvous of two spacecraft (not as simple as one might think, 1965, *Gemini 76*, US), the first docking of two spacecraft (1966, *Gemini 8* with the *Agena 8*, US), the first manned missions beyond low Earth orbit (1968, *Apollo 8*, US), landing people on the Moon (1969, *Apollo 11*, US), and numerous others.

Achievements of the 1970s include the first robotic rover (1970, *Lunakhod 1*, USSR), the first planetary orbiter (1971, Mars, *Mariner 9*, US), the first spacecraft to pass through the asteroid belt (1972, *Pioneer 10* and *11*, also the first two spacecraft to reach interstellar space, US), the first pictures from the surface of Venus (1975, *Venera 9* and *10*, USSR), the first joint docking of spacecraft built by two different countries (1975, *Apollo-Soyuz*, US-USSR), and the first pictures from the surface of Mars (1976, *Viking 1*, US).

The USSR's *Salyut* was the first space station. It was launched in 1971, occupied for 22 days and deorbited after 175 days. *Skylab*, launched in 1973 was the first US space station and hosted three crews during 1973 and 1974 for 28, 59, and 84 days. *Skylab* demonstrated the abilities and value of a laboratory in orbit, returning a wealth of scientific information ranging from Earth-observation to biomedical to materials science to astronomical, and all that with less than 6 months total of operation.

http://www.hq.nasa.gov/office/pao/History/SP-4209/ch11-3.htm

[56] NASA Scientific and Technical Information http://www.sti.nasa.gov/tto/

[57] Thomas Edison received a patent for his electric incandescent lamp on January 27, 1880.

Bratcher, Kathryn Anne, "'Went to the Exposition Tonight': Louisville's 1883 Southern Exposition," *The Filson Newsmagazine*, Vol. 4, No. 1, The Filson Historical Society, http://www.filsonhistorical.org/news_v4n1_ expo.html

[58] In 1977 smallpox was the first disease to be eradicated worldwide, after an extensive vaccination program. The reduction in polio is due to the development of vaccines by Jonas Salk in 1955 and by Albert Sabin in 1957. As of 2006, polio is on the verge of being the second disease to be eradicated. http://www.polio.info/polio-eradication/front/index.jsp?&siteCode=POLIO

Direct coal use by industry is greatly reduced, although most of the electricity for industry is generated by burning coal and other fossil fuels.

[59] "President Reagan's remarks following the loss of the Space Shuttle *Challenger* and her crew," January 28, 1986, http://www.nasa.gov/audience/formedia/speeches/reagan_challenger.html

Stuckey, Mary E., *Slipping the Surly Bonds: Reagan's Challenger Address*, Texas A&M University Press, 2006.

[60] "Space Shuttle Oversight Hearing," Committee on Science, Commerce, and Transportation, United States Senate, January 22, 1987, http://www.gpoaccess.gov/challenger/72_663.pdf

"Report of the Presidential Commission on the Space Shuttle *Challenger* Accident," Washington, D.C., June 6, 1986, http://history.nasa.gov/rogersrep/genindex.htm

[61] Cochran, E. B., *Planning Production Costs: Using the Improvement Curve*, Chandler Publishing Company, San Francisco, 1968.

[62] "Contractor's Final Documentation Report First Incremental Buy for Increment II of the Space Transportation System (STS)," Morton Thiokol, TWR-50181, December 1989, NASA CR-183855, 1991. http://ntrs.nasa.gov/archive/nasa/casi.ntrs.nasa.gov/19910001721_1991001721.pdf

"Probe Begins on Morton-Thiokol Fire," *The Victoria Advocate*, December 31, 1987 http://news.google.com/newspapers?nid=861&dat=19871230&id=Ty0PAAAAIBAJ&sjid=ZYUDAAAAIBAJ&pg=4580,142116

[63] "The History of Challenger Center for Space Science Education," http://www.challenger.org/about/history/challenger_center.cfm

[64] Udall, Stewart L., "Let's stop sleepwalking through history," *Los Angeles Times*, December 12, 2004. In a 1955 statement to the U.S. Congress, physicist John von Neumann claimed that nuclear power would be "too cheap to meter." http://www.dawn.com/2004/12/12/int13.htm

[65] Organ, Michael, "*Metropolis* Film Archive 2007,"
http://www.michaelorgan.org.au/metroa.htm,

Edelstein, David, "Radiant City - The Timely Return of Fritz Lang's *Metropolis*, One of the Greatest Ballets Ever Put on Film," *Slate*, September 18, 2002. http://www.slate.com/?id=2071036

[66] Neuharth, Al, "Space race to Mars between U.S., China?," *USA Today*, 2002, http://www.usatoday.com/news/opinion/columnists/neuharth/neu020.htm

Wheeler, Larry, "U.S. Losing Unofficial Space Race, Congressmen Say," Space.com, 31 March 2006, http://www.space.com/news/ft_060331_nasa_china_congress.html

Whittington, Mark, "The coming space race with China," *The Space Review*, June 23, 2003, http://www.thespacereview.com/article/28/1

"Moon Shots: China, Japan In '07; U.S., India in '08," *Moon Daily*, Beijing (XNA), January 4, 2007, http://www.moondaily.com/reports/Moon_Shots_China_Japan_In_2007_US_India_In_2008_999.html

David, Leonard, "Year of the Moon: China, Japan Ready Lunar Probes for '07," Space.com, 02 January 2007, http://www.space.com/news/070102_asia_moonprobes.html

Bodeen, Christopher, "Chinese cheer space walk, look to future," Associated Press, September 28, 2008.

Berger, Brian, "NASA Stresses Global Participation in New Lunar Plan," *Space News*, 13 December 2006, http://www.space.com/news/061213_moonbase_international.html

Gugliotta, Guy, "Space: The Next Generation," *National Geographic*, Vol. 212, No. 4, pp. 106-125, October 2007.

Report of the Space Shuttle Columbia Accident Investigation Board, 2003, http://caib.nasa.gov/

Griffin, Michael, "Remarks to the Space Flight Suppliers Conference," March 4, 2008 http://www.spaceref.com/news/viewsr.html?pid=27222

[67] "Marshall's Role in *Skylab*," MSFC History Office, undated, http://history.msfc.nasa.gov/skylab/skylab_planning.html; MSFC was responsible for system engineering, integrated development and verification, and cluster integration; MSC (now JSC) for mission analysis and operations; KSC for launching,

Newkirk, Roland W., and Ivan D. Ertel, *Skylab: A Chronology*, NASA SP-4011, 1977, http://history.nasa.gov/SP-4011/cover.htm

[68] McKibben, Bill, "Carbon's New Math," *National Geographic*, pp. 32-37, October 2007.

[69] Coleman, Joseph, "Study: $45 trillion needed to combat warming," Associated Press, June 6, 2008, http://news.yahoo.com/s/ap/20080606/ap_on_sc/japan_iea_climate_change

Doyle, Alister, "World needs to axe greenhouse gases by 80 pct: report," Reuters, April 19, 2007, http://www.reuters.com/article/topNews/idUSL194440620070419

"G8 plans 50% reduction in greenhouse gases," Reuters, 8 July 2008, http://www.independent.co.uk/environment/climate-change/g8-plans-50-reduction-in-greenhouse-gases-862281.html

Jorgensen, Rene, "Important step towards Copenhagen '09," Greenvoice.com, September 1, 2008, http://blog.greenvoice.com/?tag=g8-summit

Goering, Laurie, "U.N. Panel warns of 'abrupt' effects of climate change," *Chicago Tribune*, November 18, 2007

"Global Warming: The Perfect Storm," presentation to the Royal College of Physicians, London, United Kingdom, 29 January 2008, http://www.columbia.edu/~jeh1/2008/RoyalCollPhyscns_20080129.pdf

McKibben, Bill, "Remember This: 350 Parts Per Million," *Washington Post*, December 28, 2007, http://www.washingtonpost.com/wp-dyn/content/article/2007/12/27/AR2007122701942.html

350.org, http://www.350.org

Hansen, James, et. Al., "Target Atmospheric CO_2: Where Should Humanity Aim?," April 7, 2008, http://www.columbia.edu/~jeh1/2008/TargetCO2_20080407.pdf

Romm, Joseph, "Hansen (et al) must read: Get back to 350 ppm or risk an ice-free planet," *Climate Progress*, March 17, 2008, http://climateprogress.org/2008/03/17/hansen-et-al-must-read-back-to-350-ppm-or-risk-an-ice-free-planet/

Wald, Matthew, "For Carbon Emissions, a Goal of Less than Zero," *The New York Times*, March 26, 2008, http://www.nytimes.com/2008/03/26/business/businessspecial2/26negative.html?_r=1&oref=slogin

Fletcher, Kenneth R., "Interview: Wallace Broecker," *Smithsonian*, Vol. 39, No. 3, p. 23, June 2008.

Broecker, Wallace S., and Robert Kunzig, *Fixing Climate: What Past Climate Changes Reveal about the Current Threat—and How to Counter It*, Hill and Wang, 2008.

[70] Nordhaus, Ted, and Michael Shellenberger, *Break Through: From the Death of Environmentalism to the Politics of Possibility*, Houghton Mifflin, 2007

[71] McDonough, William, and Michael Braungart, *Cradle to Cradle: Remaking the Way We Make Things*, North Point Press, 2002

Brown, Lester R., *Plan B 3.0: Mobilizing to Save Civilization*, Earth Policy Institute, W. W. Norton & Company, 2008 http://www.earth-policy.org/index.php?/books/pb/pb_table_of_contents

[72] "End Poverty 2015, UN Millenium Development Goals," http://www.un.org/millenniumgoals/

[73] Hogan, Thor, *Mars Wars: The Rise and Fall of the Space Exploration Initiative*, NASA, Washington, D.C., SP-2007-4110, May 2007. http://history.nasa.gov/sp4410.pdf

Dinerman, Taylor, "Fannie Mae, Freddie Mac, and lessons for space commercialization," *The Space Review*, October 13, 2008, http://www.thespacereview.com/article/1229/1

Hedman, Eric R., "How to know when an engineering project is failing," *The Space Review*, June 30. 2008. http://www.thespacereview.com/article/1158/1

Lafleur, Claude, "Space exploration at a crossroad, Part 1: lessons to be learned," *The Space Review*, October 6, 2008, http://www.thespacereview.com/article/1225/1

Lafleur, Claude, "Space exploration at a crossroad, Part 2: What should we do?," *The Space Review*, October 13, 2008, http://www.thespacereview.com/article/1227/1

[74] "Presidential Directive on National Space Policy," February 11, 1988 http://www.hq.nasa.gov/office/pao/History/policy88.html

Hudgins, Edward L., "Next Step for U.S. Space Policy," The Heritage Foundation, February 23, 1988, http://www.heritage.org/Research/SmartGrowth/bu70.cfm

"Space's Uncertain Future," editorial in *The Huntsville Times*, Huntsville, Alabama, February 15, 1988.

[75] "Speech by OSTP Director John Marburger to the 44th Robert H. Goddard Memorial Symposium," Goddard Space Flight Center, Greenbelt, Maryland, March 15, 2006. http://www.spaceref.com/news/viewsr.html?pid=19999, OSTP = Office of Science and Technology Programs

[76] Bush, George H. W., "Remarks on the 20th Anniversary of the *Apollo 11* Moon Landing," July 20, 1989, http://www.presidency.ucsb.edu/ws/index.php?pid=17321

Several reports considered the future of the space program:

Pioneering the Space Frontier: An Exciting Vision of Our Next Fifty Years in Space, The Report of the National Commission on Space, Bantam Books, Inc., New York, 1986, http://history.nasa.gov/painerep/begin.html

"Book Review - Pioneering the Space Frontier," *Sky and Telescope*, V. 72, No. 4, p. 361, October 1986.) In June 1987 the Office of Exploration was established at NASA, "to fund, lead, and coordinate studies examining potential approaches to human exploration of the solar system." ("Report of the 90-Day Study on Human Exploration of the Moon and Mars," National Aeronautics and Space Administration, NASA-TM-102999, November 1989, p. 1-1, http://history.nasa.gov/90_day_study.pdf

"Human Exploration of Space: A Review of NASA's 90-Day Study and Alternatives," National Academy Press, Washington, D.C., 1990, http://www.nap.edu/openbook.php?record_id=10985&page=R1

In August 1987 "Leadership and America's Future in Space," by Sally Ride, was released, pointing out the need for a long-range direction and a clear strategy for

NASA. (Ride, Sally K., *Leadership and America's Future in Space*, NASA, August 1987, http://history.nasa.gov/riderep/main.pdf)

The 1988 Presidential Directive on National Space Policy was "to expand human presence and activity beyond Earth orbit into the solar system." (Fact Sheet, "Presidential Directive on National Space Policy," February 11, 1988, http://www.hq.nasa.gov/office/pao/History/policy88.html)

"Space Exploration Initiative (SEI)," GlobalSecurity.org, http://www.globalsecurity.org/space/systems/sei.htm

[77] One technology development program was the Space Station Evolution Program. In 1990 it was expected that Space Station Freedom would have a crew of 8 astronauts initially, and support as many as 30 during on-orbit preparations for Lunar/Mars missions.

Wieland, Paul O., William R. Humphries, "ECLSS Development for Future Space Missions," AIAA-1990-3728, AIAA Space Programs and Technologies National Conference, Huntsville, AL, American Institute of Aeronautics and Astronautics, September 25-28, 1990.

Another technology development program was the Pathfinder program, oriented toward developing the technologies and acquiring information needed to return people to the Moon and send them to Mars.

Early human space missions (Mercury, Gemini, Soyuz, Skylab, etc.) used "open-loop" life support systems, with separated CO_2 and wastewater vented to space. For long-duration missions, especially those away from LEO where resupply is feasible, this approach is prohibitively expensive. "Closing the loop" by reusing the separated waste products, by recovering O_2 and potable water, greatly reduces the total mass required for long-duration missions.

[78] The SBIR program is designed to provide maximum benefit for minimal cost and is divided into three phases. Phase I awards provided $50,000 (in 2007 $100,000) for 6 months of effort to verify the feasibility of a concept. Phase II awards provided up to $250,000 (in 2007 $600,000) for 2 years to build a working prototype. http://sbir.gsfc.nasa.gov/SBIR/SBIR.html

[79] The Microlith concept was originally developed in 1991 for the EPA to improve automotive catalytic converters to meet ultra-low emission vehicle standards in a smaller and lighter device that requires less precious-metal catalyst than standard catalytic oxidizers. (Tony Anderson, PCI, 12/20/07)

http://epa.gov/ncer/sbir/success/pdf/precision_combustion_success.pdf

Precision Combustion, Inc. http://www.precision-combustion.com/

Roychoudhury, S., D. Walsh, and J. Perry, "Microlith Based Sorber for Removal of Environmental Contaminants," ICES 2004-01-2442, Society of Automotive Engineers, International Conference on Environmental Systems, http://www.precision-combustion.com/2004-01-2442%20Sorber.pdf

Initially, the Microlith unit was expected to be used on board the *ISS*, but due to program changes, it was not. If it had been installed in the *ISS*, the projected savings were valued at $31 million over the 15-year life. http://techtran.msfc.nasa.gov/SBIR/converter.html

A Technology Maturity Scale is shown in: Wieland, Paul, *Designing For Human Presence in Space*, NASA RP-1324, page 62, NASA, 1994. http://ntrs.nasa.gov/search.jsp?R=535455&id=5&qs=Ne%3D25%26Ns%3DArchiveN ame%257C0%26N%3D4294652904%2B282

Roychoudhury, Subir, Marco Castaldi, Maxim Lyubovsky, Rene LaPierre, and Shabbir Ahmed, "Microlith catalytic reactors for reforming iso-octane-based fuels into hydrogen," *Journal of Power Sources*, Volume 152, pp. 75-86, December 1, 2005.

[80] The goal was to land people on Mars by 2015. Personal log book entry dated 10/4/89.

[81] Miller, R.L. and C.H. Ward, "Algal bioregenerative systems," In: E. Kammermeyer (ed.) *Atmosphere in space cabins and closed environments*. Appleton-Century-Croft Pub., New York. 1966

Taub, R.B., "Closed ecological systems," In: R.F. Johnston, P.W. Frank, and C.D. Michener (eds.) *Ann. Rev. Ecology Systematics*. Annual Reviews Inc., Palo Alto, CA. pp 139-160. 1974

Tadros, Mahasin G., "Characterization of *Spirulina* Biomass for CELSS Diet Potential," Alabama A&M University, Normal, AL, NASA contractor report NCC 2-501, October 1988, http://ntrs.nasa.gov/archive/nasa/casi.ntrs.nasa.gov/19890016190_ 1989016190.pdf

Gitelson, I.I., G. M. Lisovsky, and R. D. MacElroy, "Manmade Closed Ecological Systems: Bioregenerative Life Support Systems for Use in Space and Other Hosts," Earth Space Institute, ESI Book Series, CRC Press, 2003, http://books.google.com/books?id=vxnV-2k6z2wC

Nelson, Mark, "Workshop Introduction," Third International Workshop on Closed Ecological Systems, April 24-27, 1992, Krasnoyarsk, Siberia. http://www.biospheres.com/hist3rdwkshopintro.html

BIOS-3, http://en.wikipedia.org/wiki/BIOS-3

[82] *Biosphere 2*, http://www.biospherics.org

[83] Chandler, David L., "10 Lessons from *Biosphere 2*," *Wired*, Issue 12.12, December 2004, http://www.wired.com/wired/archive/12.12/biosphere.html

articles in *Habitation* (formerly *Life Support and Biosphere Science*), http://www.cognizantcommunicatino.com/filecabinet/Habitation/hab.html

Biosphere 2, http://www.infoplease.com/ce6/sci/A0807629.html

Marries, Dan, "A Look Back on the *Biosphere II* Experiment 12 Years Later," Kold News, October 10, 2005, http://www.kold.com/Global/story.asp?S=3902230)

[84] Janik, Daniel Scott, "President's Greeting," Institute for Advanced Systems in Life Support (IASLS) Newsletter, 1(1), 1991, http://www.getcited.org/pub/103404038

Crump, William J., "From the Editor in Chief," *The Journal of Life Support and Biospheric Science*, Cognizant Communications, Volume 1, Number 1, 1994.

[85] "Inside *Biosphere 2*," Columbia University, http://www.columbia.edu/cu/ 21stC/issue-2.1/specmain.htm

"Ocean acidification due to increasing atmospheric carbon dioxide," The Royal Society, London, UK, ISBN 0 85403 617 2, June 2005, http://royalsociety.org/displaypagedoc.asp?id=13539

Nagel, Jennifer, "Projected Atmospheric Change Threatens Coral Reefs, *Biosphere 2* Study Reveals," Columbia University Record, vol. 24, no. 21, April 23, 1999, http://www.columbia.edu/cu/newrec/2421/tmpl/story.10.html

"Increasing Carbon Dioxide Threatens Coral Reefs," *Science Daily*, May 18, 2000, http://www.sciencedaily.com/releases/2000/05/000516114559.htm

"Marine Calcifiers in a High CO_2 Ocean," *Science*, May 23, 2008.

Harris, William C., and Lisa J. Graumlich, "*Biosphere 2*: sustainable research for a sustainable planet," 21stC, http://www.columbia.edu/cu/21stC/issue-4.1/harris.html

Griffin, Kevin L., et al, "Plant growth in elevated CO_2 alters mitochondrial number and chloroplast fine structure," Proceedings of the National Academy of Sciences, vol. 98, no. 5, February 27, 2001. http://www.pnas.org/cgi/content/full/98/5/2473?ck=nck

Lorenz, Edward N., "Predictability: Does the Flap of a Butterfly's Wings in Brazil Set Off a Tornado in Texas?," American Association for the Advancement of Science, December 1972, Summary of theory at: http://www.viewzfromscience.com/documents/webpages/chaos_p3.html, http://whatis.techtarget.com/definition/0,,sid9_gci759332_top1,00.html, http://www.fortunecity.com/emachines/e11/86/beffect.html

Gleick, James, *Chaos: Making a New Science*, Penguin Books, 1988.

[86] Jones, Harry, "The Systems Engineering Process for Human Support Technology Development," ICES-2005-157093, 35th International Conference on Environmental Systems, Rome, Italy, July 11-14, 2005. http://ntrs.nasa.gov/search.jsp?R=159606& id=7&qs=No%3D0%26N% 3D4294728181

"NASA Space Flight Program and Project Management Requirements," NASA Procedural Requirements, NPR 7120.5D, March 6, 2007, http://spacecraft.ssl.umd.edu/design_lib/NASA-7120.5D.pdf

[87] "Wilhelm Albert," http://www.britannica.com/eb/topic-12706/Wilhelm-Albert

"Brooklyn Bridge," http://www.inventionfactory.com/history/RHAbridg/bb.html

"Roebling and the Brooklyn Bridge," http://memory.loc.gov/ammem/today/jun12.html

"World's Longest Bridge Spans," http://www.tkk.fi/Units/Bridge/longspan.html,

"List of Longest Suspension Bridge Spans,"
http://en.wikipedia.org/wiki/List_of_largest_suspension_bridges

[88] "The Next Incandescent," *Energy & Power Management*, Vol. 32, No. 4, page 23, a BNP publication, April 2007.

[89] Kim, Hyeon-Hye, Raymond M. Wheeler, John C. Sager, Neil C. Yorio, and Gregory. D. Goins, "Light-Emitting Diodes as an Illumination Source for Plants: A Review of Research at Kennedy Space Center," *Habitation*, Vol. 10, pp. 71-78, 2005 Cognizant Comm. Corp.,
http://www.cognizantcommunication.com/filecabinet/Habitation/hab10abs2.html

Ilieva, Iliyana, Tania Ivanova, Yordan Naydenov, Ivan Dandolov, and Detelin Stefanov, "Plant experiments with light-emitting diode module in *Svet* space greenhouse," 37th COSPAR Scientific Assembly. 13-20 July 2008, Montréal, Canada, p.1316, http://adsabs.harvard.edu/abs/2008cosp...37.1316I

"First commercial plant growth experiment from Wisconsin team gets started on Space Station," press release, Marshall Space Flight Center, NASA, May 10, 2001, http://www.spaceref.com/news/viewpr.html?pid=4811

Tsao, Jeffrey, "Roadmap projects significant LED penetration of lighting market by 2010," *Laser Focus World*, May 1, 2003, http://www.laserfocusworld.com/articles/article_display.html?id=177632

Teichner, Martha, "Let There Be LEDs!," *CBS News Sunday Morning*, May 18, 2008, http://www.cbsnews.com/stories/2008/05/18/sunday/main4105020.shtml

Eisenberg, Anne, "In Pursuit of Perfect TV Color, With L.E.D.'s and Lasers," *The New York Times*, June 24, 2007. http://www.nytimes.com/2007/06/24/business/yourmoney/24novel.html

Long, Colleen, "New Year light ball will have eco-bulbs," The Associated Press, December 31, 2007.

"New York City Uses New Ball Technology," *Pollution Engineering*, January 2008, p. 12.

"LED Flashlights," Equipped to Survive, http://www.equipped.com/led_lights.htm

"Wal-Mart Uses GE LED Refrigerated Display Lighting," *Energy & Power Management*, Vol. 32, No. 7, p. 25, July 2007.

Krakow, Gary, "LED bulbs: a bright idea?," MSNBC, September 19, 2005, http://www.msnbc.msn.com/id/9399209

Solid-State Lighting, Department of Energy, http://www.netl.doe.gov/ssl/

"The Great Internet Light Bulb Book, Part I,"
http://freespace.virgin.net/tom.baldwin/bulbguide.html

[90] "Wind May Generate 30 Percent of Electricity by 2030 - Study," Reuters News Service, September 21, 2006, http://www.planetark.com/dailynewsstory.cfm/newsid/38191/story.htm;

Svenvold, Mark, "Wind-Power Politics," *The New York Times*, September 12, 2008, http://www.nytimes.com/2008/09/14/magazine/14wind-t.html?partner=rssyahoo& emc=rss

"A Framework for Offshore Wind Energy Development in the United States," U.S. Department of Energy, Massachusetts Institute of Technology, General Electric, September 2005, http://www.mtpc.org/offshore/final_09_20.pdf;

Zajac, Jennifer, "Offshore wind: Gaining momentum or getting blown off?," SNL *Energy Generation Markets Week*, Vol. 5, Issue 8, February 21, 2006

"Grid energy storage," Wikipedia, viewed September 13, 2008, http://en.wikipedia.org/wiki/Grid_energy_storage

Baxter, Richard, Sr., "Energy Storage - Supporting Greater Wind Energy Usage," Energy Pulse, Energy Central Network, December 16, 2005, http://www.energypulse.net/censters/article/article_display.cfm?a_id=1164

[91] Cha, Ariana Eunjung, "Solar Energy Firms Leave Waste Behind in China," *Washington Post*, March 9, 2008, http://www.washingtonpost.com/wp-dyn/content/article/2008/03/08/AR2008030802595_pf.html

Liu, Yingling, "The Dirty Side of a 'Green' Industry," The Worldwatch Institute, March 14, 2008, http://www.worldwatch.org/node/41

Masamitsu, Emily, "Startup Makes Cheap Solar Film Cells ... With an Inkjet Printer," *Popular Mechanics*, March 6, 2008. http://www.popularmechanics.com/science/earth/4253464.html?series=15

Vidal, John, "Solar energy 'revolution' brings green power closer," *The Guardian*, 29 December 2007, http://www.guardian.co.uk/environment/2007/dec/29/solarpower.renewableenergy

Blake, Mariah, "The Rooftop Revolution," *Washington Monthly*, March/April 2009, http://www.washingtonmonthly.com/features/2009/0903.blake.html

Goh, Chiatzun, and Michael D. McGehee, "Organic Semiconductors for Low-Cost Solar Cells," National Academy of Engineering, The National Academies, Volume 35, Number 4, Winter 2005, http://www.nae.edu/NAE/bridgecom.nsf/weblinks/MKEZ-6LULZ6?OpenDocument

Zwiebel, Ken, James Mason, and Zasilis Fthenakis, "A Solar Grand Plan," *Scientific American*, January 2008, p. 64-73

Makhijani, Arjun, "Carbon-Free and Nuclear-Free: A Roadmap for U.S. Energy Policy," Nuclear Policy Research Institute and the Institute for Energy and Environmental Research, IEER Press and RDR Books, 2007. http://www.ieer.org/carbonfree/index.html

Frosch, Dan, "Citing Need for Assessments, U.S. Freezes Solar Energy Projects," *The New York Times*, June 27, 2008, http://www.nytimes.com/2008/06/27/us/27solar.html

[92] Keith, David W., Joseph F. DeCarolis, David C. Denkenberger, Donald H. Lenschow, Sergey L. Malyshev, Stephen Pacala, and Philip J. Rasch, "The influence of large-scale wind power on global climate," PNAS, November 16, 2004, vol. 101, no. 46, 16115-16120, http://www.pnas.org/cgi/doi/10.1073/pnas.0406930101

Wald, Matthew L., "Wind Energy Bumps Into Power Grid's Limits," *The New York Times*, August 26, 2008, http://www.nytimes.com/2008/08/27/business/27grid.html

[93] Tester, Jefferson W., Elisabeth M. Drake, Michael J. Driscoll, Michael W. Golay, William A. Peters, *Sustainable Energy: Choosing Among Options*, The MIT Press, 2005.

"Geothermal Energy Gathering Steam," Associated Press, October 7, 2008, http://www.cbsnews.com/stories/2008/10/07/tech/main4507263.shtml?tag=lowerContent;homeSectionBlock205

"Geothermal Energy Set to Double Across Western States," Environment News Service, January 14, 2008, http://www.ens-newswire.com/ens/ jan2008/2008-01-14-094.asp ,

Geothermal Energy Association, http://www.geo-energy.org/

"Study: geothermal energy could meet 10% of future US electricity needs," Renewable Energy Today, January 23, 2007, http://findarticles.com/p/articles/mi_m0OXD/is_2007_Jan_23/ai_n17218424

Choi, Charles Q., "MIT study: Get more energy from Earth's heat," msnbc, January 22, 2007, http://www.msnbc.msn.com/id/16755646/ Geothermal power plants could provide 10% of our electrical needs by 2050. Concerns relate to water usage and the potential for earthquakes when water is injected.

"The Future of Geothermal Energy," Massachusetts Institute of Technology, 2006, http://geothermal.inel.gov/publications/future_of_geothermal_energy.pdf, Geothermal energy environmental impacts: While the overall environmental impact of geothermal energy is low, "Most of the potentially important environmental impacts of geothermal power plant development are associated with ground water use and contamination, and with related concerns about land subsidence and induced seismicity as a result of water injection and production into and out of a fractured reservoir formation. Issues of air pollution, noise, safety, and land use also merit consideration."

"Renewables 2007: Global Status Report," REN21, Renewable Energy Policy Network for the 21st Century, http://www.ren21.net, http://www.worldwatch.org/files/pdf/renewables2007.pdf

[94] Some components of a nuclear fusion power plant, however, would become radioactive, requiring careful disposal. http://www.iter.org/a/index_nav_4.htm

Schneider, Ursula, "Fusion: Energy of the Future," WorldAtom Staff Report, IAEA Physics Section, International Atomic Energy Agency, August 1, 2001, http://www.iaea.org/NewsCenter/News/2001/08012001_news02.shtml

"Fusion In Our Future," *ScienceDaily*, September 10, 1998, http://www.sciencedaily.com/releases/1998/09/980910074918.htm

Sofge, Erik, "MIT Fights for Clean Power With Holy Grail of Fusion in Reach," *Popular Mechanics*, February 25, 2008, http://www.popularmechanics.com/science/research/4251982.html

Gellerman, Bruce, "Some Like It Hot....," *Living On Earth*, February 24, 2006, http://www.loe.org/shows/shows.htm?programID=06-P13-00008

ITER was formerly the International Thermonuclear Experimental Reactor. http://www.iter.org/

Kestenbaum, David, "Researchers Seek to Recreate Fusion Power," National Public Radio, Climate Connections: Solutions, August 20, 2007, http://www.npr.org/templates/story/story.php?storyId=13746131

Clery, Daniel, "Design Changes Will Increase ITER Reactor's Cost," *Science*, Vol. 320, no. 5882, p. 1405, 13 June 2008, The U.S. share of the ITER cost is about 9%, or $1.122 billion. http://fire.pppl.gov/iter_cost_science_061308.pdf

Brumfiel, Geoff, "Fusion reactor faces cost hike," *Nature*, Vol 453, number 829, 12 June 2008, http://www.nature.com/news/2008/080612/full/453829a.html

"Nuclear fusion must be worth the gamble," Editorial, *New Scientist*, 7 June 2006, http://environment.newscientist.com/channel/earth/energy-fuels/mg19025543.300-editorial-nuclear-fusion-must-be-worth-the-gamble.html

"Department of Energy Assessment of the ITER Project Cost Estimate," November 2002, http://www.ofes.fusion.doe.gov/News/ITERCostReport.pdf

"U.S. Signs International Fusion Energy Agreement," U.S. Department of Energy, November 21, 2006, http://www.energy.gov/news/4486.html, ITER, http://www.iter.org/

[95] "Net-Zero Energy Commercial Building Initiative," Building Technologies Program, Energy Efficiency & Renewable Energy, U.S. Department of Energy, http://www1.eere.energy.gov/buildings/commercial_initiative/

Energy Star, U.S. Department of Energy, http://www.energystar.gov/

Zero Energy Commercial Buildings Consortium, http://zeroenergycbc.org/

Torcellini, P., S. Pless, M. Deru, and D. Crawley, "Zero Energy Buildings: A Critical Look at the Definition," ACEEE Summer Study, Pacific Grove, CA, August 14-18, 2006, Conference Paper NREL/CP-550-39833, June 2006 http://www.nrel.gov/docs/fy06osti/39833.pdf

[96] Alternative sources of energy and fuel:

"The National Methane Hydrates R&D Program," National Energy Technology Laboratory, U.S. Department of Energy, http://www.netl.doe.gov/technologies/oil-gas/FutureSupply/MethaneHydrat es/maincontent.htm

Seibert, Michael, Paul King, Liping Zhang, Lauren Mets, and Maria Ghirardi, "Molecular Engineering of Algal H_2 Production," Proceedings of the 2002 U.S. DOE

Hydrogen Program Review, NREL/CP-610-32405,
http://www1.eere.energy.gov/hydrogenandfuelcells/pdfs/32405a3.pdf

"Rare Microorganism That Produces Hydrogen May Be Key To Tomorrow's Hydrogen Economy," *ScienceDaily*, July 8, 2008,
http://www.sciencedaily.com/releases/2008/07/080707192643.htm

"Algae oil promises truly green fuel," *NewScientist*, 10 June 2008,
http://environment.newscientist.com/channel/earth/energy-fuels/mg19826595.900-algae-oil-promises-truly-green-fuel.html

Ghirardi, Maria L., and Michael Seibert, "Algal Hydrogen Photoproduction," National Renewable Energy Laboratory, Golden, CO, Program Review Meeting, May 19-22, 2003. http://www1.eere.energy.gov/hydrogenandfuelcells/pdfs/32405a3.pdf

"E. coli a future source of energy?," physorg.com, January 29, 2008,
http://www.physorg.com/news120846806.html

Hallenbeck, Patrick C., and John R. Benemann, "Biological hydrogen production; fundamentals and limiting processes," *International Journal of Hydrogen Energy*, Vol. 27, Numbers 11-12, p. 1185-1193, November/December 2002.

Patel, Sonal, "Renewables, A new wave: Ocean power," *Power*, May 2008
http://www.powermag.com/ArchivedArticleDisplay2.aspx?a=48-F_Renew&y=2008&m=may

Chapa, Jorge, "Underwater Ocean Turbines Will Generate Renewable Energy," *Inhabitat*, December 10, 2007, http://www.inhabitat.com/2007/12/10/underwater-power-generating-ocean-turbines/

"PG&E mounts tidal power project," *Power*, Vol. 151, No. 8, p. 6, August 2007,
www.powermag.com

Oceanlinx Limited - http://www.oceanlinx.com/, DEXA Wave Energy - http://www.dexawave.com/?gclid=CIKF7NedqZMCFRIvxwodFBVXoA

McKenna, Phil, "Nanomaterial turns radiation directly into electricity," *NewScientist*, 27 March 2008, http://environment.newscientist.com/channel/earth/energy-fuels/dn13545-nanomaterial-turns-radiation-directly-into-electricity.html

"New Device Turns Waste Heat into Electricity," LiveScience, 04 June 2007,
http://www.livescience.com/technology/070604_sound_electricity.html

"Feeling the Heat: Berkeley Researchers Make Thermoelectric Breakthrough in Silicon Nanowires," Physorg.com, January 10, 2008,
http://www.physorg.com/news119201015.html

Bullis, Kevin, "Hot Advance for Thermoelectrics: Cheap organic molecules could more efficientyl convert wasteheat into electricity," *Technology Review*, MIT, February 22, 2007. http://www.technologyreview.com/Energy/18211.

McKenna, Phil, "Reincarnated material turns waste heat into power," *NewScientist*, 20 March 2008, http://environment.newscientist.com/channel/earth/energy-fuels/dn13512-reincarnated-material-turns-waste-heat-into-power.html

"'Cold Fusion' Rebirth? Symposium Explores Low Energy Nuclear Reactions," *ScienceDaily*, March 30, 2007, http://www.sciencedaily.com/releases/2007/03/070329095612.htm

Gellerman, Bruce, "Cold Fusion: A Heated History," *Living On Earth*, February 24, 2006, http://www.loe.org/shows/shows.htm?programID=06-P13-00008

Van Noorden, Richard, "Cold Fusion Back on the menu," Chemistry World, 22 March 2007, http://www.rsc.org/chemistryworld/News/2007/March/22030701.asp

"Flying wind farms," *The Economist*, 3 April 2007, http://www.economist.com/science/tq/displaystory.cfm?story_id=8952080

Behar, Michael, "Windmills in the Sky," *Popular Science*, 21 November 2005, http://www.popsci.com/scitech/article/2005-11/windmills-sky

Sky Windpower Corporation, http://www.skywindpower.com/ww/index.htm

Levesque, Tylene, "Flying Wind Turbines," *Inhabitat*, July 17, 2007, http://www.inhabitat.com/2007/07/17/flying-wind-turbines/

[97] "My New Hexa-Pent Dome Designed for You to Live In," *Popular Science*, May 1972.

Fuller, R. Buckminster, "The Year 2000," Speech at San Jose State College, March 1966, http://www.lauralee.com/news/bucky2000.htm

[98] The Defense Advanced Research Projects Agency (DARPA) was established in 1958 to provide ongoing Department of Defense support for technology advances. "The Government Role in Civilian Technology: Building a New Alliance," National Academy Press, Washington, D.C., 1992, http://books.nap.edu/openbook.php?record_id=1998&page=62

[99] Livingston, David, "Is space exploration worth the cost?," *The Space Review*, January 21, 2008, http://www.thespacereview.com/article/1040/1

The Centennial Challenges program is a part of the Innovative Partnerships Program, that offers prizes to advance the technologies needed for a variety of aerospace challenges. http://www.nasa.gov/offices/ipp/innovation_incubator/centennial_challenges/index.html

[100] Fuller, Buckminster, *Earth, Inc.,* Anchor Books, 1973. http://www.nous.org.uk/EarthInc.html

[101] Choi, Charles Q., "Electric Vehicles Could Strain Water Supplies," Live Science, March 10, 2008, http://news.yahoo.com/s/livescience/electricvehiclescouldstrainwatersupplies

Barta, Patrick, "Amid Water Shortage, Australia Looks to the Sea," *The Wall Street Journal*, March 11, 2008, http://online.wsj.com/article/SB120518234721525073.html

[102] Hansen, James R., "Technology and the History of Aeronautics: An Essay," U.S. Centennial of Flight Commission, The First Law of the History of Technology of Melvin C. Kranzberg. undated, http://www.centennialofflight.gov/essay/Evolution_of_Technology/Tech-OV1.htm

[103] Revkin, Andrew C., "A Shift in the Debate Over Global Warming," *The New York Times*, April 6, 2008, http://www.nytimes.com/2008/04/06/weekinreview/06revkin.html

von Hippel, Frank N., "Rethinging Nuclear Fuel Recycling," *Scientific American*, May 2008

[104] Hartmann, William K., "The Next 25 Years in Space," chapter in Ordway, Frederick I., III, and Randy Liebermann, eds., *Blueprint for Space: Science Fiction to Science Fact*, Smithsonian Institution Press, 1992

[105] Clarke, Arthur C., *The Promise of Space*, Harper and Row, 1968

[106] Logsdon, John M., "Human Space Flight and National Power," *High Frontier*, March 2007, p. 11-13, http://www.gwu.edu/~spi/

U.S.-Soviet Cooperation in Space, U.S. Congress, Office of Technology Assessment, OTA-TM-STI-27, Washington, DC July 1985. http://www.fas.org/ota/reports/8533.pdf

[107] O'Brien, Miles, "NASA Symposium: Risk and Exploration," September 27, 2004, http://www.risksymposium.arc.nasa.gov/docs/Transcript1.pdf, http://74.125.47.132/search?q=cache:http://www.risksymposium.arc.nasa.gov/docs/Transcript2.pdf, http://www.spaceref.com/news/viewsr.html?pid=19994

Foust, Jeff, "A skeptic's guide to space exploration," *The Space Review*, June 30, 2008, http://www.thespacereview.com/article/1160/1, Review of a speech at International Space Development Conference in 2008. According to Neil deGrasse Tyson the reasons for exploration are: war, praise of royalty or deity, and the promise of economic return.

Relatively few people went to California during the Gold Rush. The 1850 census showed only 92,597 people (not counting the native peoples), and the 1860 census 379,994. In 1890 the population was 1,213,398. (compare with population increases in Iowa, etc in endnote 32)

http://www.city-data.com/states/California-Population.html

Lewis, John S., *Mining The Sky: Untold Riches From the Asteroids, Comets, and Planets*, Basic Books, 1997.

Wingo, Dennis, *Moonrush: Improving Life on Earth with the Moon's Resources*, Apogee Books Space Series 43, 2004.

[108] Oberg, James, "The secret formula for going to the moon: Fear played a role in 1960s, and may do so again," MSNBC, July 19, 2004, http://www.msnbc.msn.com/id/5380736/

Mosher, Dave, "Crater Could Solve 1908 Tunguska Meteor Mystery," *Live Science*, June 26, 2007. http://www.livescience.com/space/ scienceastronomy/070626_st_tunguska_crater.html

Burrows William E., *The Survival Imperative: Using Space to Protect Earth*, Forge Books, 2006.

[109] Olsen, Stefanie, "NASA budget emphasizes space exploration," cnet News.com, February 6, 2006, http://news.cnet.com/NASA-budget-emphasizes-space-exploration/ 2100-11397_3-6035753.html

[110] Reagan, Ronald, "Remarks at Edwards Air Force Base, California, on Completion of the Fourth Mission of the Space Shuttle Columbia," July 4, 1982, http://www.reagan.utexas.edu/archives/speeches/1982/70482a.htm

[111] "AirLaunch LLC Selected for Contract Continuance by DARPA," The Free Library by Farlex, November 3, 2005, http://www.thefreelibrary.com/AirLaunch+LLC+Selected+for+Contract+Continuance +by+DARPA.-a0138287057   In 2003, nine companies were awarded contracts by DARPA to develop initial designs for a small rocket.

[112] "Jet Contrails Alter Average Daily Temperature Range," *ScienceDaily*, August 8, 2002, http://www.sciencedaily.com/releases/2002/08/020808075457.htm

Stenger, Richard, "9/11 study: Air traffic affects climate," CNN.com, August 8, 2002, http://archives.cnn.com/2002/TECH/science/08/07/contrails.climate/

Travis, David J., Andrew M. Carleton, and Ryan G. Lauritsen, "Contrails reduce daily temperature range," *Nature*, Vol. 418, 8 August 2002, http://facstaff.uww.edu/travisd/pdf/jetcontrailsrecentresearch.pdf

Tyson, Peter, "The Contrail Effect," NOVA, April 2006 http://www.pbs.org/wgbh/nova/sun/contrail.html

"Clouds Caused By Aircraft Exhaust May Warm The U.S. Climate," *ScienceDaily*, April 28, 2004, http://www.sciencedaily.com/releases/2004/04/040428061056.htm

"NASA Scientists Use Empty Skies To Study Climate Change," *ScienceDaily*, May 22, 2002, http://www.sciencedaily.com/releases/2002/05/020522074456.htm

[113] Britt, Robert Roy, "Rocket Exhaust Leaves Mark Above Earth" Space.com, 10 June 2003, http://www.space.com/scienceastronomy/ shuttle_clouds_030610.html

"NRL Study Finds Shuttle Exhaust Is Source Of Mysterious Clouds In Antarctica," ScienceDaily, July 7, 2005, http://www.sciencedaily.com/releases/2005/07/ 050707055950.htm

"Study Finds Space Shuttle Exhaust Creates Night-Shining Clouds," NASA, June 3, 2003. http://www.nasa.gov/centers/goddard/news/topstory/2003/0522shuttleshine.html

Sebacher, Daniel I., Richard J. Bendura, and Gerald L. Gregory, "Hydrogen Chloride Measurements in the Space Shuttle Exhaust Cloud—First Launch, April 12, 1981,"

Journal of Spacecraft and Rockets, 1982, 0022-4650 vol. 19 no. 4 (366-370), AIAA 82-4176, American Institute of Aeronautics and Astronautics, http://www.aiaa.org/content.cfm?pageid=406&gTable=japapcrimportPre97& gID= 62266

Stevens, Michael H., Christoph R. Englert, and Jorg Gumbel, "OH observations of space shuttle exhaust," *Geophysical Research Letters*, Vol. 29, No. 10, 1378, American Geophysical Union, 21 May 2002 http://www.agu.org/pubs/crossref/ 2002/2002GL015079.shtml

Merle, Renae, "Additional $340 Million Is Proposed For Rockets," *Washington Post*, February 8, 2005, http://www.washingtonpost.com/wp-dyn/articles/A6463-2005Feb7.html

Brady, B.B., E.W. Fournier, L.R. Martin, and R.B. Cohen, "Stratospheric Ozone Reactive Chemicals Generated by Space Launches Worldwide," Aerospace Corporation, Defense Technical Information Center, Accession Number: ADA289852, 1 November 1994, http://stinet.dtic.mil/oai/oai?verb=getRecord&metadataPrefix= html&identifier=ADA289852

[114] The primary method for producing industrial quantities of H_2 is by steam reforming of methane, that also produces CO_2 in a two-step reaction process: (1) $CH_4 + H_2O$ -> $3H_2 + CO$ (2) $CO + H_2O$ -> $CO_2 + H_2$

Other methods are being developed, but are not yet in wide use.

Palmer, Jason, "Solar cell speeds hydrogen production," *NewScientist.com*, 18 February 2008, http://environment.newscientist.com/channel/earth/energy-fuels/ dn13344-solar-cell-speeds-hydrogen-production.html

[115] Kleiner, Kurt, "Exotic nitrogen could offer safe rocket fuel," *NewScientist*, 28 May 2008, http://environment.newscientist.com/channel/earth/energy-fuels/ mg19826584.300-exotic-nitrogen-could-offer- safe-rocket-fuel.html

Eremets, Mikhail I., Alexander G. Gavriliuk, Ivan A. Trojan, Dymitro A. Dzivenko, and Reinhard Boehler, "Single-bonded cubic form of nitrogen," *Nature Materials*, Vol. 3, pp. 558-563, August 2004, published online 4 July 2004, http://www.nature.com/nmat/journal/v3/n8/abs/nmat1146.html;jsessionid =C65337E2E3EDEE057BD7CF2BD3238A22

Goho, Alexandra, "Nitrogen Power: New crystal packs a lot of punch," *Science News*, July 17, 2004, http://www.phschool.com/science/science_news/articles/ nitrogen_power.html, http://www.accessmylibrary.com/article-1G1-120188920/ nitrogen-power-new-crystal.html

[116.] JP Aerospace, http://www.jpaerospace.com/

Powell, John M., *Floating to Space*, Apogee Books, May 2008.

For the Airship-to-Orbit, another balloon (the Ascender) would carry people and cargo from the ground to the platform, from where the orbital balloon ship would be "launched."

Faust, Jeff, "Floating to space," *The Space Review*, June 1, 2004, http://www.thespacereview.com/article/151/1

Boyle, Alan, "Airship groomed for flight to edge of space," MSNBC, May 21, 2004, http://www.msnbc.msn.com/id/5025388/

"Advanced Electric Propulsion," University of Washington, 28 July 2004, http://www.ess.washington.edu/Space/propulsion.html

"Electric Spacecraft Propulsion," ESA, http://sci.esa.int/science-e/www/object/index.cfm?fobjectid=34201&fbodylongid=1537

"Hall Effect Thrusters," ESA, http://sci.esa.int/science-e/www/object/index.cfm?fobjectid=34201&fbodylongid=1538

"Magnetoplasmadynamic thrusters," NASA Glenn Research Center, http://www.nasa.gov/centers/glenn/about/fs22grc.html

[117] Saravanan, D., "Spider Silk - Structure, Properties and Spinning," *Journal of Textile and Apparel*, Technology and Management, Volume 5, Issue 1, Winter 2006, http://www.tx.ncsu.edu/jtatm/volume5issue1/Articles/Saravanan/Saravanan_Full_170_05.pdf

Miller, Steve, "Spinning Superstuff," Houghton Mifflin Science, http://www.eduplace.com/science/hmsc/6/e/cricket/cktcontent_6e143.shtml

Vergano, Dan, "Success! Scientists spin spider silk," *USA Today*, January 20, 2002, http://www.usatoday.com/news/science/biology/2002-01-21-spider-silk.htm

Thiel, B.L., C. Viney, and Lynn W. Jelinski, "β Sheets and Spider Silk," *Science*, Vol. 273, no. 5281, 13 September 1996, http://www.sciencemag.org/cgi/content/citation/273/5281/1477e

[118] Dupont website, http://www2.dupont.com/Kevlar/en_US/index.html

"Technical Guide, Kevlar® Aramid Fiber," http://www2.dupont.com/Kevlar/en_US/assets/downloads/KEVLAR_Technical_Guide.pdf

[119] "Diamond," Encarta, http://encarta.msn.com/encyclopedia_761557986/diamond.html

Smalley, Richard, Harold Kroto, and Robert Curl, "C_{60} Buckminsterfullerene," *Nature*, 14 November 1985.

Kroto was a chemist at the University of Sussex, England. He was studying the carbon molecules in space and went to Rice University to use Smalleys equipment to recreate the conditions in the atmosphere of a red giant star to study the carbon clusters produced.

Greenfieldboyce, Nell, "'Buckyball' Nobel Laureate Richard Smalley Dies," *All Things Considered*, National Public Radio, October 31, 2005, http://www.npr.org/templates/story/story.php?storyId=4983474

Aldersey-Williams, Hugh, *The Most Beautiful Molecule: The Discovery of the Buckyball*, John Wiley & Sons, 1995

Farnsworth, Martha, Maclovio Fernandez, and Luca Sabbatini, *Buckyballs: Their history and discovery*, Rice University, 2005, http://cnx.org/content/m14355/latest/

Iijima, Sumio, Masako Yudasaka, and Fumiyuki Nigey, "Carbon Nanotube Technology," NEC Corporation, undated, http://www.nec.co.jp/techrep/en/journal/g07/n01/070110.html

Yakobson, Boris, and Richard Smalley, "Fullerene Nanotubes: $C_{1,000,000}$ and Beyond," *American Scientist*, July/August 1997, http://www.americanscientist.org/issues/feature/fullerene-nanotubes-c1000000-and-beyond/1

[120] Yakobson, Boris, and Richard Smalley, "Fullerene Nanotubes: $C_{1,000,000}$ and Beyond," *American Scientist*, July/August 1997, http://www.americanscientist.org/issues/feature/fullerene-nanotubes-c1000000-and-beyond/1

Winters, Jeffrey, "Wonder Cloth," *Mechanical Engineering*, Vol. 128, No. 4,, p. 34-35. April 2006 http://www.memagazine.org/backissues/membersonly/apr06/features/wondercl/wondercl.html

"'Buckypaper' Hype Could Soon Be Reality," Associated Press, October 17, 2008, http://www.cbsnews.com/stories/2008/10/17/tech/main4528442.shtml?tag=lowerContent;homeSectionBlock205

Vigolo, Brigitte, Alain Penicaud, Claude Coulon, Cedric Sauder, Rene Pailler, Catherine Journet, Patrick Bernier, and Phillipe Poulin, "Macroscopic Fibers and Ribbons of Oriented Carbon Nanotubes," *Science*, Vol. 290, no. 5495, pp. 1331-1334, 17 November 2000, http://www.sciencemag.org/cgi/content/ abstract/290/5495/1331

[121] Clarke, Arthur C., "Extra-Terrestrial Relays," *Wireless World*, October 1945, http://lakdiva.org/clarke/1945ww/

Pearson, Jerome, "Konstantin Tsiolkovsky and the Origin of the Space Elevator," IAF-97-IAA.2.1.09, International Astronautical Federation, 48th IAF Congress, Torino, Italy, October 6-10, 1997.

"Space Elevator," NOVA Science Now, Public Broadcast System, http://www.pbs.org/wgbh/nova/sciencenow/3401/02.html

Artsutanov, Y., "Into the Cosmos by Electric Rocket," *Komsomolskaya Pravda*, 31 July 1960, http://www.star-tech-inc.com/id4.html

Isaacs, John D., Allyn C. Vine, Hugh Bradner, and George E. Bachus, "Satellite Elongation into a True 'Sky-Hook'," *Science*, Vol. 151, no. 3711, 11 February 1966, http://www.sciencemag.org/cgi/content/abstract/151/3711/682

Pearson, J., "The Orbital Tower: A Spacecraft Launcher Using the Earth's Rotational Energy," *Acta Astronautica* 2, 785-799, 1975.

"Space Elevators and other Advanced Concepts by Jerome Pearson," http://www.star-tech-inc.com/spaceelevator.html

[122] The cable "tenacity" is the strength (GPa) divided by the density (g/cc), http://www.spaceward.org/elevator-when

The tension along the length of the cable (or ribbon) will vary with distance from the Earth, due to decreasing gravity and increasing centrifugal force. As a result, the ribbon can be thinned by tapering, which will reduce the overall mass.

Yakobson, Boris, and Richard Smalley, "Fullerene Nanotubes: $C_{1,000,000}$ and Beyond," *American Scientist*, July/August 1997, http://www.americanscientist.org/issues/feature/fullerene-nanotubes-c1000000-and-bcyond/1

Nugent, Tom, "Space Elevator Ribbon Mass and Taper Ratio," LiftPort Inc., November 16, 2005, http://www.liftport.com/papers/2005Nov_LP-Ribbon_ Mass.pdf

"Going Up? Forget Shuttles, Next Time Take the Space Elevator," *Newsweek*, September 30, 2003

David, Leonard, "High hopes for space elevator," MSNBC, September 17, 2003, http://www.msnbc.msn.com/id/3077701/

Cowen, Ron, "Ribbon to the Stars: Pushing the space elevator closer to reality," *Science News*, October 5, 2002, Vol. 162, No. 14, p. 218. http://www.sciencenews.org/articles/20021005/bob9.asp

Marks, Paul, "Space elevators needed for space solar power?," *New Scientist,* January 5, 2009, http://www.newscientist.com/blogs/shortsharpscience/2009/01/space-elevators-needed-for-spa.html

"The Great Space Elevator," The Unmuseum, http://www.unmuseum.org/spaceelevator.htm

"The Space Elevator Reference," SpaceRef, http://www.spaceelevator.com/

"NASA Plans Elevators to Space," April 17, 2002, http://usgovinfo.about.com/library/weekly/aa041702a.html

Liftport Group, http://www.liftport.com/research2.php

[123] Clarke, Arthur C., "The Space Elevator: 'Thought' Experiment or Key to the Universe?," Address to the XXXth International Astronautical Congress, Munich, 20 September 1979, Advances in Earth Oriented Applied Space Technologies, Vol. 1, pp. 39 to 48, Pergamon Press Ltd., 1981, http://www.islandone.org/LEOBiblio/CLARK3.HTM

Melloan, Jim, "Going Up?," Inc. magazine, June 2004, http://www.inc.com/magazine/20040601/entreinspace.html

The Liftport Group is aiming for 2031 to have an operational space elevator. http://www.liftport.com/

DeVault, Ryan Christopher, "Building an Elevator to Space—Closer to Reality Than You May Assume!," Associated Content, July 20, 2008, http://www.associatedcontent.com/article/893083/building_an_elevator_to_space_closer.html

"A Chance to Imagine the Future," ESA News, European Space Agency, 13 December 2004, http://www.esa.int/esaCP/SEMOCAXJD1E_index_ 0.html

Gilbertson, Roger G., "Riding a Beam of Light: NASA's First Space Elevator Competition Proves Highly Challenging," SPACE.com, 24 October 2005, http://www.space.com/businesstechnology/051024_spaceelevator_challenge.html

The Spaceward Foundation Elevator 2010 Competition, http://www.spaceward.org/elevator2010

Olsen, Stefanie, "Space elevator isn't going anywhere yet," CNET News.com, October 22, 2007, http://news.cnet.com/Space-elevator-isnt-going-anywhere-yet/2100-11397_3-6214726.html

[124] Edwards, Bradley C., "The Space Elevator NIAC Phase II Final Report," NASA Institute for Advanced Concepts, March 1, 2003, http://www.niac.usra.edu/files/studies/final_report/521Edwards.pdf

Edwards, Bradley C., "The NIAC Space Elevator Program," NASA Institute for Advanced Concepts, March 1, 2003 http://www.spaceelevator.com/docs/NIACpaper.pdf

[125] Young, Kelly, "Space elevators: 'First floor, deadly radiation'," New Scientist, 13 November 2006, http://space.newscientist.com/article/dn10520-space-elevators-first-floor-deadly-radiation.html

"Protecting the Space Shuttle from Meteoroids and Orbital Debris," National Academy Press, Washington, D.C., 1997, http://www.nap.edu/openbook.php?record_id=5958. A paint flake chipped a window on Columbia during mission STS-73 in 1995 and over 80 Shuttle windows have had to be replaced due to debris impacts.

"Swerve Left To Avoid That Satellite," Space Mart, July 11, 2008, http://www.spacemart.com/reports/Swerve_Left_To_Avoid_That_Satellite_999.html

Broad, William J., "Orbiting Junk, Once a Nuisance, Is Now a Threat," The New York Times, February 6, 2007, http://www.nytimes.com/2007/02/06/science/space/06orbi.html

Atkinson, Nancy, "Space Debris Illustrated: The Problem in Pictures," Universe Today, April 11, 2008, http://www.universetoday.com/2008/04/11/space-debris-illustrated-the-problem-in-pictures/

NASA Orbital Debris Program Office, http://orbitaldebris.jsc.nasa.gov/

"Orbital Debris: A Technical Assessment," Committee on Space Debris, National Research Council, The National Academies Press, 1995, http://www.nap.edu/catalog.php?record_id=4765

[126] Crouch, D.S., and M.M. Vignoli, "Shuttle tethered satellite system development program," AIAA-1984-1106, Space Systems Technology Conference, Costa Mesa, CA, June 5-7, 1984, Technical Papers (A84-34004 15-12), New York, American Institute of Aeronautics and Astronautics, 1984, p. 21-31.

Lavoie, Anthony R., "Tethered Satellite System (TSS-1R)-Post Flight (STS-75) Engineering Performance Report," NASA Marshall Space Flight Center, report JA-

2422, August 1996, http://ntrs.nasa.gov/archive/nasa/casi.ntrs.nasa.gov/ 20010022502_2001031372.pdf

[127] Pennicott, Katie, "Nanotubes are the new superconductors," *PhysicsWeb*, June 28, 2001, http://physicsworld.com/cws/article/news/2652

Tang, Z.K, Lingyun Shang, N. Wang, X.X. Zhang, G.H. Wen, G.D. Li, J.N. Wang, C.T. Chan, and Ping Sheng, "Superconductivity in 4 Angstrom Single-Walled Carbon Nanotubes," *Science*, Vol. 292, no. 5526, pp. 2462-2465, 29 June 2001, http://www.sciencemag.org/cgi/content/abstract/292/5526/2462.

Cho, Adrian, "Nanotubes hint at room temperature superconductivity," *New Scientist*, 28 November 2001, http://www.newscientist.com/article/dn1618-nanotubes-hint-at-room-temperature-superconductivity.html

Cleuziou, J.-P., W. Wernsdorfer, V. Bouchiat, T. Ondarcuhu, and M. Monthioux, "Carbon nanotube superconducting quantum interference device," *Nature Nanotechnology* 1, 53-59 (2006), published online 4 October 2006, http://www.nature.com/nnano/journal/v1/n1/full/nnano.2006.54.html

Man, H.T., "Carbon Nanotube-based Superconducting and Ferromagnetic Hybrid Systems," Masters Thesis, TU Delft, 12 September 2006, http://www.tudelft.nl/live/pagina.jsp?id=6c664112-dfa5-4710-a0ef-ad8a86aa3774&lang=en

Smitherman, D.V., "Space Elevators: An Advanced Earth-Space Infrastructure for the New Millennium," NASA Marshall Space Flight Center, NASA/CP-2000-210429, August 2000.

"Space Elevator History," Star, Inc., http://www.star-tech-inc.com/id4.html

Vaughn, Jason A., Leslie Curtis, Brian E. Gilchrist, Sven Bilen, and Enrico Lorenzini, "Review of the ProSEDS Electrodynamic Tether Mission Development," AIAA, ASME, SAE, ASEE 40th Joint Propulsion Conference, Fort Lauderdale, FL, 11-14 July 2004, http://ntrs.nasa.gov/search.jsp?R=421801&id=4&qs=Ne%3D25%26Ns%3DArchiveN ame%257C0%26N%3D4294807149%2B127, The experiment was scheduled to fly in 2003, but the loss of *Columbia* led to cancellation of the test, even though it was to fly on a Delta II rocket.

Tethers Unlimited Inc., http://www.tethers.com/, Several applications for space tethers are described including orbital debris removal.

[128] Stanley-Robinson, Kim, *Red Mars*, Spectra, 1993

[129] "Overview of the Electric Grid," U.S. Department of Energy, http://www.energetics.com/gridworks/grid.html

U.S.. Energy Information Agency, *International Energy Annual 2006*, World Electricity Data, http://www.eia.doe.gov/emeu/iea/elec.html

U.S. Department of Energy, Office of Electricity Delivery & Energy Reliability, http://www.oe.energy.gov/

Lobsenz, George, "PacifiCorp Seeks FERC Aid On Nation's Biggest Grid Expansion Project," The Energy Daily, July 8, 2008, http://www.theenergydaily.com/publications/ed/1053.html

Pentland, William, and Brian Wingfield, "Same Wires, More Power," Forbes, July 7, 2008, http://www.forbes.com/business/2008/07/03/energy-efficiency-electricity-biz-energy_cx_bw_wp_0707efficiency_grid.html

"Commissioning Of World's First Superconductor Power Transmission Cable System Celebrated," ElectricNet, July 2, 2008, http://www.electricnet.com/article.mvc/Commissioning-Of-Worlds-First-Superconductor-0001?VNETCOOKIE=NO

American Superconductor Corporation, http://www.amsc.com/

Haught, Debbie, "Status and Future Outlook of Superconductivity Program at DOE," Office of Electricity Delivery and Energy Reliability, September 17, 2007, http://www.oe.energy.gov/DocumentsandMedia/Haught_MSandT__9_17_07.pdf

Forsyth, E.B., "Superconducting power transmission systems-the past and possibly the future," Superconducting Science Technology, Vol. 6, no. 10, pp. 699-714, October 1993, http://www.iop.org/EJ/abstract/0953-2048/6/10/001

Larbalestier, David, Richard D. Blaugher, Robert E. Schwall, Robert S. Sokolowski, Masaki Suenaga, and Jeffrey O. Willis, "Power Applications of Superconductivity in Japan and Germany," World Technology Evaluation Center, September 1997, http://www.wtec.org/loyola/scpa/toc.htm

Gelsi, Steve, "Power firms grasp net tech for aging grid," MarketWatch, July 11, 2008, http://www.marketwatch.com/news/story/power-firms-grasp-new-technology/story.aspx?guid=%7B3BB486EE-6B51-4B5D-9E91-0099ED4ED291%7D

Heger, Monica, "Superconductors Enter Commercial Utility Service," IEEE Spectrum Online, Institute of Electrical and Electronics Engineers, 2 July 2008, http://spectrum.ieee.org/energy/the-smarter-grid/superconductors-enter-commercial-utility-service

[130] Clarke, Arthur C., "The Space Elevator: 'Thought' Experiment or Key to the Universe?," Address to the XXXth International Astronautical Congress, Munich, 20 September 1979, Advances in Earth Oriented Applied Space Technologies, Vol. 1, pp. 39 to 48, Pergamon Press Ltd., 1981, http://www.islandone.org/LEOBiblio/CLARK1.HTM

[131] Weiler, Edward J., "Testimony of Edward J. Weiler Given at a Senate Science, Technology, and Space Hearing on Space Propulsion," June 3, 2003, http://www.spaceref.com/news/viewsr.html?pid=9355

"SMART-1: The First Spacecraft Of The Future," Science Daily, September 25, 2003, http://www.sciencedaily.com/releases/2003/09/030925070219.htm

Hoverstein, Paul, "Deep Space 1 sets record with ion propulsion system," Space.com, 17 August 2000, http://www.space.com/scienceastronomy/solarsystem/deepspace_propulsion_000816.html

Dawn Mission to Ceres and Vesta, Jet Propulsion Laboratory, NASA, http://www.jpl.nasa.gov/news/features.cfm?feature=1468

Reisz Engineers, http://www.reiszengineers.com/

Carreau, Mark, "Fast Trip," *Aviation Week & Space Technology*, pp. 63-65, August 10, 2009

[132] Clarke, Arthur C., *Project Solar Sail*, ROC, Penguin Books, 1990.

Oberg, James, "Solar-sail mission reflects past and future," MSNBC, June 20, 2005, http://www.msnbc.msn.com/id/8291710/

Vulpetti, Giovanni, Les Johnson, and Gregory L. Matloff, *Solar Sails: A Novel Approach to Interplanetary Travel*, Copernicus Books, 2008.

Foeust, Jeff, "Review: Solar Sails," *The Space Review*, October 6, 2008, http://www.thespacereview.com/article/1223/1

Benford, Gregory, and Paul Nissenson, "Reducing solar sail escape times from Earth orbit using beamed energy," *Acta Astronautica*, Vol. 58, No. 4, February 2006, http://www.sciencedirect.com/science?_ob=ArticleURL&_udi=B6V1N-4HSY4H3-1&_user=10&_rdoc=1&_fmt=&_orig=search&_sort=d&view=c&_acct=C000050221&_version=1&_urlVersion=0&_userid=10&md5=ff4eabad9546acd31ac9fad1a23a3fb9

"NASA MESSENGER Sets Record for Accuracy of Planetary Flyby," Press Release, Johns Hopkins University, http://www.spaceref.com/news/ viewpr.html?pid=26655

"Setting Sail for the Stars," *Science@NASA*, June 28, 2000, http://science.nasa.gov/headlines/y2000/ast28jun_1m.htm

Coulter, Dauna, "A Brief History of Solar Sails," Physorg.com, August 1, 2008, http://www.physorg.com/news136810834.html

[133] "Japan Deploys Solar Sail Film in Space," SpaceRef.com, August 10, 2004, http://www.spaceref.com/news/viewpr.html?pid=14782

Japanese Aerospace Exploration Agency http://www.isas.ac.jp/e/snews/2004/0809.shtml

Leonard, David, "Planetary Society's Cosmos 1 Solar Sail Ready for Flight," Space.com, 9 November 2004, http://www.space.com/missionlaunches/cosmos-1_update_041109.html

"NASA to Attempt Historic Solar Sail Deployment," *Science@NASA*, June 26, 2008. http://science.nasa.gov/headlines/y2008/26jun_nanosaild.htm, Note: The spacecraft was lost during launch when the rocket malfunctioned.

Hsu, Jeremy, "First Solar Sail Might Fly Soon," Space.com, 12 August 2009, http://www.space.com/businesstechnology/090812-tw-solar-sail-new-mission.html

[134] "Methane Blast," *Science@NASA*, 04 May 2007, http://science.nasa.gov/headlines/y2007/04may_methaneblast.htm

[135] Dewar, James A., *To The Ends of the Solar System: The Story of the Nuclear Rocket*, second edition, Apogee Books, 2007.

[136] David, Leonard, "NASA's Nuclear Prometheus Project Viewed as Major Paradigm Shift," Space.com, 07 February 2003, http://www.space.com/businesstechnology/technology/prometheus_030207.html

"Project Prometheus: Jupiter Icy Moons Orbiter Fact Sheet," NASA, February 2003, http://nssdcftp.gsfc.nasa.gov/miscellaneous/jupiter/JIMO_Background/JIMO.pdf

"Ion Thrusters Propel NASA Into Future," May 27, 2005, http://www.nasa.gov/vision/universe/features/nep_prometheus.html

Berger, Brian, "NASA 2006 Budget Presented: Hubble, Nuclear Initiative Suffer," Space News, 07 February 2005, http://www.space.com/news/nasa_budget_050207.html

[137] "Odyssey: A Program for Human Exploration of Space," The California Institute of Technology, March 10, 2004, http://www.its.caltech.edu/~epstein/odyssey_executive_summary_web_files/frame.htm

[138] Wieland, P.O., M.C. Roman, and L. Miller, *Living Together in Space: The International Space Station Internal Active Thermal Control System Issues and Solutions—Sustaining Engineering Activities at the Marshall Space Flight Center, 1998 to 2005*, NASA TM-2007-214964, National Aeronautics and Space Administration, Washington, D.C., Marshall Space Flight Center, AL, 2007. http://ntrs.nasa.gov/search.jsp?R=643765&id=1&qs=Ntt%3Dwieland%26Ntk%3DAuthorList%26Ntx%3Dmode%2520matchall%26N%3D4294967039%2B42%26Ns%3DHarvestDate%257c1

[139] Gilmore, David G., *Spacecraft Thermal Control Handbook: Fundamental Technologies*, American Institute of Aeronautics and Astronautics, 2002. http://books.google.com/books?id=-GYG lwG8PkUC

"Ask an Astronomer," *Cool Cosmos*, California Institute of Technology, http://coolcosmos.ipac.caltech.edu/cosmic_kids/AskKids/moontemp.shtml

[140] Griffin, Michael, "House Committee on Science Holds a Hearing on the Future of NASA," congressional testimony, June 28, 2005, http://www.nasa.gov/pdf/119619main_Griffin_Hil_testimony_062805.pdf

Malik, Tariq, "Report: Space Radiation a Serious Concern for NASA's Exploration Vision," Space.com, 23 October 2006, http://www.space.com/news/061023_space_radiation.html

Roop, Lee, and Shelby G. Spires, "NASA forms Plan B team," *The Huntsville Times*, March 5, 2010

[141] "Space Radiation Threats To Astronauts Addressed In Federal Research Study," ScienceDaily, October 30, 2006, http://www.sciencedaily.com/releases/2006/10/061025184743.htm

"Managing Space Radiation Risk in the New Era of Space Exploration," Committee on the Evaluation of Radiation Shielding for Space Exploration, National Research Council, 2008, http://books.nap.edu/catalog.php?record_id=12045&utm_medium=etmail&utm_source=National%20Academies%20Press&utm_campaign=New+from+NAP+4.07.08-May-24-2008&utm_content=Downloader&utm_term=

"NASA Works on Radiation Protection Shield," Associated Press, 01 December 2003, http://www.space.com/scienceastronomy/nasa_radiation_031201.html,

Thibeault, S.A., et al., "Development of Improved This Polymer Films for Space Structures and Radiation Shielding," 2001, http://lowdose.tricity.wsu.edu/2001mtg/abstracts/thibeault.htm

Townsend, L.W., "Critical analysis of active shielding methods for space radiation protection," Aerospace Conference, 2005 IEEE, March 5-12, 2005, pp. 724-730. http://ieeexplore.ieee.org/Xplore/login.jsp?url=/iel5/10432/33126/01559364.pdf?isnumber=33126&prod=CNF&arnumber=1559364&arSt=724&ared=730&arAuthor=Townsend%2C+L.W.

Berardelli, Phil, "Solar Storm! Shields Up!," ScienceNOW Daily News, 04 November 2008, http://sciencenow.sciencemag.org/cgi/content/full/ 2008/1104/1

"Shields For The Starship *Enterprise*: A Reality?," *ScienceDaily*, April 19, 2007, http://www.sciencedaily.com/releases/2007/04/070419113601.htm

Malik, Tariq, "Lunar Shields: Radiation Protection for Moon-Based Astronauts," Space.com, 12 January 2005, http://www.space.com/businesstechnology/lunarshield_techwed_050112.html

Malik, Tariq, "Shields Up! New Radiation Protection for Spacecraft and Astronauts," Space.com, 27 May 2004, http://www.space.com/businesstechnology/technology/rad_shield_040527.html

Chaikin, Andrew, "Radiation Dangers for Mars Astronauts Downgraded," *Sky & Telescope*, July 23, 2003, http://www.skyandtelescope.com/ news/3307401.html

[142] Solar Terrestrial Relations Observatory, NASA Goddard Space Flight Center, http://stereo.gsfc.nasa.gov/mission/mission.shtml

Goudarzi, Sara, "STEREO Ready to Take on the Sun," Space.com, 17 August 2006, http://www.space.com/missionlaunches/060817_stereo_launch.html

[143] "2001 *Mars Odyssey*," Jet Propulsion Laboratory, http://mars.jpl.nasa.gov/odyssey/overview/

Britt, Robert Roy, "*Mars Odyssey* Shows Intense, But Manageable Radiation Risk for Astronauts," Space.com, 13 March 2003, http://www.space.com/missionlaunches/odyssey_radiation_030313.html

"NASA's Space Radiation Laboratory," NASA, http://spaceflight.nasa.gov/shuttle/support/researching/radiation/

Choi, Charles Q., "Space Radiation Too Deadly For Mars Mission," SPACE.com, 31 March 2008, http://www.space.com/missionlaunches/080331-radiation-shielding.html

[144] Street, Kenneth W., Jr., Christian Schrader, and Doug Rickman, "Some Expected Characteristics of Lunar Dust: A Geological View Applied to Engineering," The Geological Society of America, Paper No. 345-5, 2008 Joint Meeting of the Geological Society of America, Houston, TX, 9 October 2008, http://gsa.confex.com/gsa/2008AM/finalprogram/abstract_150730.htm

Christoffersen, R., J.F. Lindsay, and J.A. Lawrence, "Lunar Dust Effects on Spacesuit Systems: Insights from the Apollo Spacesuits," The Geological Society of America, Paper No. 345-8, Houston, TX, 9 October 2008, http://gsa.confex.com/gsa/2008AM/finalprogram/abstract_150328.htm

Gugliotta, Guy, "Can We Survive on the Moon?," Discover, March 21, 2007, http://discovermagazine.com/2007/mar/can-we-survive-on-the-moon

"2008 Regolith Excavation Challenge," California Space Education and Workforce Institute, http://regolith.csewi.org/

Taylor, Lawrence A., and Thomas T. Meeks, "Microwave Sintering of Lunar Soil: Properties, Theory, and Practice," J. Aerosp. Engrg. Volume 18, Issue 3, pp. 188-196 (July 2005), http://scitation.aip.org/getabs/servlet/GetabsServlet?prog=normal&id=JAEEEZ000018000003000188000001&idtype=cvips&gifs=yes

[145] Kaufman, Marc, "Microbes May Threaten Lengthy Spaceflights," Washington Post, December 10, 2007, http://www.washingtonpost.com/wp-dyn/content/article/2007/12/09/AR2007120900665.html

Britt, Robert Roy, "Surviving Space: Risks to Humans on the Moon and Mars," Space.com, 20 January 2004, http://www.space.com/scienceastronomy/mars_dangers_040120.html

Schneider, Mike, "What if an astronaut snaps in space?," Associated Press, Huntsville Times, February 24, 2007.

Piquepaille, Roland, "Self-Healing Computers for NASA Spacecraft," April 25, 2008, http://blogs.zdnet.com/emergingtech/?p=903

"Exploration Technology Development," NASA Exploration Directorate, http://www.nasa.gov/exploration/acd/technology_dev.html

[146] "Space Station Freedom," Encyclopedia Astronautica, http://www.astronautix.com/craft/spaeedom.htm

Wheeler, Larry, "Space Station's Total Cost Remains an Elusive Number," Florida Today, 01 December 2002, http://www.space.com/missionlaunches/iss_cost_011201.html

Leary, Warren E., "Fate of Space Station Is in Doubt As All Options Exceed Cost Goals," The New York Times, June 8, 1993, http://query.nytimes.com/gst/fullpage.html?res=9F0CE7D71530F93BA35755C0A965958260

Hess, Mark, "Station Redesign Team To Submit Final Report," NASA Press Release 93-104, June 4, 1993. http://www.nasa.gov/home/hqnews/ 1993/93-104.txt

Lemonick, Michael D., "The Next Giant Leap for Mankind," *Time*, July 24, 1989, http://www.time.com/time/magazine/article/0,9171,958208-1,00.html

Dick, Steve, "Summary of Space Exploration Initiative," http://history.nasa.gov/seisummary.htm

"The Space Exploration Initiative," NASA, http://history.nasa.gov/sei.htm

[147] Smith, Marcia, "NASA's Space Station Program: Evolution and Current Status," Testimony before the House Science Committee, April 4, 2001, http://history.nasa.gov/smith.htm

"Spacecraft: Manned: *Mir*: Chronology," http://www.russianspaceweb.com/ mir_chronology.html

"U.S. Astronauts Aboard *Mir*," *Space.com*, 12 May 2000, http://www.space.com/peopleinterviews/linenger_mirsidebar_000512.html

"Biographical Data: Sergei Konstantinovich Krikalev," NASA Johnson Space Center, http://www.jsc.nasa.gov/Bios/htmlbios/krikalev.html

[148] Thirkettle, Alan, "*ISS* and its Evolution in the Framework of Space Exploration," International Cooperation for Sustainable Space Exploration, Session 3, 5 May 2005, http://esamultimedia.esa.int/docs/spineto/2005/session3/1_esa_session3.ppt

Pellerin, Cheryl, "*International Space Station* Partners Applaud Year's Achievements," America.gov, U.S. Department of State, 26 January 2007, http://www.america.gov/st/washfile-english/2007/January/20070126120256l cnirellep6.188601e-02.html

Pellerin, Cheryl, "World Space Agencies Coordinate on Future Exploration," America.gov, U.S. Department of State, 16 July 2008, http://www.america.gov/st/ space-english/2008/July/20080716155656lcnirellep0.3303034.html?CP.rss=true

Horowitz, Scott, "Future of Space Exploration Depends on International Cooperation," America.gov, U.S. Department of State, 11 October 2006, http://www.america.gov/st/ washfile-english/2006/October/ 20061011113830lcnirellep0.9637567.html

[149] Brandenburger, Adam, and Barry Nalebuff, *Co-Opetition*, Doubleday Business, 1996.

"'Coopetition' In The New Economy: Collaboration Among Competitors," Technology Project, The Progressive Policy Institute, http://www.neweconomyindex.org/section1_page07.html

Dagnino, Giovanni Battista, and Giovanna Padula, "Coopetition Strategy: A New Kind of Interfirm Dynamics For Value Creation," The European Academy of Management, Second Annual Conference, Stockholm, Sweden, 9-11 May 2002, http://www.altruists.org/static/files/CoOpetition%20Strategy.pdf

[150] "International Cooperation and Competition in Civilian Space Activities," U.S. Congress, Office of Technosogy Assessment, OTA-ISC-239, July 1985, NTIS #PB87-136842, http://www.princeton.edu/~ota/disk2/1985/8513_n.html

[151] "Early Bird: World's First Commercial Communications Satellite," Boeing, http://www.boeing.com/defense-space/space/bss/factsheets/376/earlybird/ebird.html

"Intelsat," Mission and Space Library, NASA, http://samadhi.jpl.nasa.gov/msl/Programs/intelsat.html

"Intelsat's Satellite Communication Highlights from the 60's," Intelsat, http://www.intelsat.com/about-us/history/intelsat-1960s.asp

[152] Zak, Anatoly, "Mission Possible," *Air & Space Magazine*, August/September 2008. http://www.airspacemag.com/space-exploration/Mission_Possible.html

"Russia warns of asteroid threat," *Space Daily*, September 5, 2008, http://www.spacedaily.com/reports/Russia_warns_of_asteroid_threat_999.html

[153] Virgin Galactic, http://www.virgingalactic.com/

Schreck, Adam, "Private spaceflight edges closer," Associated Press, March 4, 2010.

Boyle, Alan, "'Coopetition' reigns among spaceports," MSNBC, October 18, 2006, http://www.msnbc.msn.com/id/15320942/

[154] Pellerin, Cheryl, "World Space Agencies Coordinate on Future Exploration," America.gov, U.S. Department of State, 16 July 2008, http://www.america.gov/st/space-english/2008/July/20080716155656lc nirellep0.3303034.html?CP.rss=true

[155] "NASA and ESA complete comparative exploration architecture study," European Space Agency, Press Release, July 9, 2008, http://www.moontoday.net/news/viewpr.html?pid=25908

[156] Griffin, Michael D., "Human Space Exploration: The Next 50 Years," *Aviation Week and Space Technology*, March 14, 2007, http://aviationweek.typepad.com/space/2007/03/human_space_exp.html

[157] Rogers, Tom, "Hearing on Space Launch Initiative: Testimony by Tom Rogers," June 20, 2001. http://www.spacefuture.com/archive/hearing_on_space_launch_initiative_testimony_by_tom_rogers.shtml

Abbey, George, Neal Lane, and John Murator, "Maximizing NASA's Potential in Flight and On the Ground: Recommendations for the Next Administration," James A. Baker III Institute for Public Policy, Rice University, January 20, 2009, http://www.bakerinstitute.org/publications/SPACE-pub-ObamaTransitionAbbeyLaneMuratore-012009.pdf

[158] Rosenberg, Matt, "Eratosthenes," About.com, http://geography.about.com/od/historyofgeography/a/eratosthenes.htm

[159] Wilson, Nigel Guy, Encyclopedia of Ancient Greece, Routledge, 2007, "Eratosthenes," pp. 269-270, http://books.google.com/books?id=-aFtPdh6-

2QC&pg=PT299&lpg=PT299&dq=eratosthenes+geography&source=web&ots=vI39s KYMjn&sig=CR7keQaI_sa35CH63G7B_UF_XyQ&hl=en&sa=X&oi=book_result&r esnum=10&ct=result#PPT298,M1

[160] The International Geophysical Year, The National Academies, http://www7.nationalacademies.org/archives/igyhistory.html

"The Sunspot Cycle," Solar Physics, Marshall Space Flight Center, http://solarscience.msfc.nasa.gov/SunspotCycle.shtml

Roederer, Juan G., "Progress in Solar-Terrestrial Physics," International Council of Scientific Unions, Scientific Committee on Solar-Terrestrial Physics, Springer, p. 19, 1983

"Archives of President Dwight David Eisenhower," June 30, 1957, http://www.eisenhower.archives.gov/Research/Digital_Documents/IGY/ IGYdocuments.html

[161] "Mission to planet Earth: International Space Year, 1992," UN Chronicle, December 1992, http://findarticles.com/p/articles/mi_m1309/is_n4_v29/ai_13344181?tag= rbxcra.2.a.1 The International Space Year was first proposed in 1985 by Sen. Spark Matsunaga of Hawaii.

Gavaghan, Helen, "An expedition to Earth," NewScientist, 29 July 1989, http://www.newscientist.com/article/mg12316752.800-an-expedition-to-earth-worries-about-the-environment-arepushing-research-into-space-this-time-attention-is-not-focused-on-distantplanets-the-idea-is-to-take-a-long-hard-look-at-earth-itself-.html

"President Bush Launches International Space Year," NASA press release 92-12, January 24, 1992. http://www.nasa.gov/home/hqnews/1992/92-012.txt

[162] John Marburger, Speech at the 44th Robert H. Goddard Memorial Symposium. March 15, 2006. http://www.spaceref.com/news/viewsr.html?pid=19999

In addition to the Hubble Space Telescope, the Compton Gamma Ray Observatory, the Chandra X-ray Observatory, and the Spitzer Infrared Observatory make up NASAs Four Great Observatories. These cover a broad range of the electromagnetic spectrum. http://www.nasa.gov/audience/forstudents/postsecondary/features/F_NASA_Great_ Observatories_PS.html

"Mars Exploration Rover Mission," Jet Propulsion Laboratory, http://marsrovers.nasa.gov/home/index.html

"Phoenix Mars Mission," University of Arizona, http://phoenix.lpl.arizona.edu/

"Phoenix Mars Lander," NASA, http://www.nasa.gov/mission_pages/phoenix/main/index.html

"NASA's Mars Phoenix Lander Returns Treasure Trove for Science," Press Release, University of Arizona, June 26, 2008, http://www.spaceref.com/news/viewpr.html?pid=25794 Mars Phoenix is the first spacecraft to perform wet chemistry and the first to "bake" a soil sample of another planet to 1000°C (1800°F).

New Horizons Mission, NASA, http://www.nasa.gov/mission_pages/newhorizons/main/index.html

MESSENGER, NASA, http://messenger.jhuapl.edu/

"MESSENGER Mission to Mercury," NASA, http://www.nasa.gov/mission_pages/messenger/main/index.html

"Fermi: Gamma Ray Space Telescope," NASA, http://fermi.gsfc.nasa.gov/

"Orbiting Carbon Observatory," NASA, http://oco.jpl.nasa.gov/, The OCO will collect precise global measurements of carbon dioxide (CO_2) in the Earth's atmosphere.

"Glory," NASA, http://glory.gsfc.nasa.gov/, Glory will monitor the Earth to better understand the energy balance.

"Kepler Mission: A search for habitable planets," NASA, http://kepler.nasa.gov/

"NASA's Shuttle and Rocket Launch Schedule," NASA, http://www.nasa.gov/missions/highlights/schedule.html

"NuSTAR: Nuclear Spectroscopic Telescope Array," Jet Propulsion Laboratory, NASA, http://www.nustar.caltech.edu/

Hautaluoma, Grey, "NASA Restarts Telescope Mission to Detect Black Holes," NASA Press Release: 07-198, September 21, 2007, http://www.nasa.gov/home/hqnews/2007/sep/HQ_07198_NuSTAR.html

The James Webb Space Telescope, NASA, http://www.jwst.nasa.gov/

Kaufman, Marc, "Shooting for the Stars With the Webb Telescope," The Washington Post, February 5, 2007, http://www.washingtonpost.com/wp-dyn/content/article/2007/02/04/AR2007020400990.html

Schneider, Mike, "NASA Spacecraft to Study Solar Flares," The Associated Press, October 25, 2006, http://www.washingtonpost.com/wp-dyn/content/article/2006/10/25/AR2006102501287.html

"Lunar Crater Observation and Sensing Satellite," NASA, http://www.nasa.gov/mission_pages/LCROSS/main/index.html

Committee on the Scientific Context for Exploration of the Moon, "The Scientific Context for Exploration of the Moon: Final Report," National Academies Press, National Research Council, 2007. http://www.nap.edu/catalog.php?record_id=11954

[163] Bodeen, Christopher, "A new space race in Asia?," The Associated Press, November 2, 2007. In 2003 China launched astronauts into space. In 2007 Chang'e 1 lunar orbiter was launched. They plan a lunar lander in 2012 and a lunar sample recovery mission in 2020.

Drew, Jill, "Space Inspires Passion and Practicality in China," Washington Post, September 25, 2008, http://www.washingtonpost.com/wp-dyn/content/article/2008/09/23/AR2008092302649.html

Page, Jeremy, "India takes on old rival China in new Asian space race," *Times Online*, June 20, 2008, http://www.timesonline.co.uk/tol/news/world/asia/article4182216.ece#cid=OTC-RSS&attr=797093

"India Maps Out Manned Exploration Program," *Moon Daily*, November 25, 2008, http://www.moondaily.com/reports/India_Maps_Out_Manned_Exploration_Program_999.html

"Iran plans to launch humans into space," *NewScientist*, 21 August 2008, http://space.newscientist.com/article/dn14576-iran-plans-to-launch-humans-into-space.html?DCMP=ILC-hmts&nsref=news2_head_dn14576.

McGrath, Matt, "France plans revolution in space," BBC News, 1 July 2008, http://news.bbc.co.uk/2/hi/science/nature/7482232.stm

Davidson, Keay, "World's nations will shoot for the moon in the next decade," *San Francisco Chronicle*, March 5, 2006, http://www.sfgate.com/cgi-bin/article.cgi?f=/c/a/2006/03/05/MOON.TMP&type=science

"*SMART-1*," European Space Agency, http://smart.esa.int/science-e/www/area/index.cfm?fareaid=10

"Probe crashes on moon just as planned," MSNBC.com, September 3, 2006, http://www.msnbc.msn.com/id/14646238/

"Deep Space Probe Completes Asteroid Flyby," CBS News, September 5, 2008, http://www.cbsnews.com/stories/2008/09/05/tech/main4421818.shtml

"SELenological and ENgineering Explorer '*Kaguya*' (*SELENE*)," JAXA, http://www.jaxa.jp/projects/sat/selene/index_e.html   *SELENE* is the first Japanese mission to orbit the Moon. Launched on September 14, 2007, the mission cost $279 million. In addition to the main orbiter, two orbiting probes will be deployed, to conduct a yearlong observational mission of the composition, geography, sub-surface structure, magnetic field, and gravity field of the Moon.

Talmadge, Eric, "Japan Lunar Probe Reaches Orbit," The Associated Press, October 5, 2007.

Chang'e 1 is the first in a series of missions to the Moon that China is planning. It was launched on October 24, 2007 and cost $187 million. It will orbit the Moon for a year, to study the composition and structure of the Moon, the space environment near the Moon, and evaluate helium-3 resources.

"The *Chang'e-1* – ProjectChina's Lunar Exploration Program (II)," China National Space Administration, http://www.cnsa.gov.cn/n615709/n772514/n772543/93747.html

"*Chang'e-1* lunar probe completes 3rd orbital transfer," Xinhua, October 29, 2007. http://www.nyconsulate.prchina.org/eng/xw/t376339.htm

"*Chandrayaan*: Lunar Mission," Indian Space Research Organization, http://www.chandrayaan-i.com/

"A Giant Leap Towards The Moon," Physorg.com, July 15, 2005, http://www.physorg.com/news5203.html

"Astronaut Makes China's First Spacewalk," Associated Press, September 27, 2008, http://www.cbsnews.com/stories/2008/09/27/tech/main4482348.shtml

[164] Statement of Edward Morris at a Hearing on Space and U.S. National Power, Committee on Armed Services, Subcommittee on Strategic Forces, U.S. House of Representatives, National Oceanic and Atmospheric Administration, U. S. Department of Commerce, June 21, 2006 http://www.space.commerce.gov/library/speeches/2006-06-spacepowerhearing.shtml

[165] Henson, Robert, Ed., "Satellite Observations to Benefit Science and Society: Recommended Missions for the Next Decade," Committee on Earth Science and Applications from Space: A Community Assessment and Strategy for the Future, National Research Council, 2008, http://books.nap.edu/catalog.php?record_id=11952&utm_medium=etmail&utm_source=National%20Academies%20Press&utm_campaign=New+from+NAP+9.16.08&utm_content=web&utm_term=

Shiga, David, "NASA calls for ambitious outer solar system mission," *New Scientist*, 05 February 2008, http://space.newscientist.com/article/ dn13276-nasa-calls-for-ambitious-outer-solar-system-mission.html

Dunham, Will, "NASA eyes dark energy, outer solar system missions," Reuters, February 6, 2008, http://www.reuters.com/article/domesticNews/idUSN0628257920080206

"Solar System Missions," NASA, Jet Propulsion Laboratory, http://www.jpl.nasa.gov/solar_system/missions/future.cfm

"Solar System Exploration," NASA, http://solarsystem.nasa.gov/missions/index.cfm

Discovery Program, NASA, http://discovery.nasa.gov/

"Deep Impact: Mission to a Comet," NASA, http://www.nasa.gov/mission_pages/deepimpact/main/index.html

"Deep Impact Legacy Site," NASA, http://solarsystem.nasa.gov/deepimpact/index.cfm

Chang, Alicia, "Comet-buster passes Earth on new mission," The Associated Press, January 2, 2008

Chang, Alicia, "Deep Impact spacecraft zips past Earth," The Associated Press, USAToday, January 1, 2008, http://www.usatoday.com/tech/science/2008-01-01-441539569_x.htm

Spires, Shelby G., "Marshall probe set to explore asteroids," *The Huntsville Times*, September 26, 2007. The Dawn spacecraft was launched in 2007 on a 15 year mission to explore Vesta and Ceres, two of the solar system's largest asteroids, at a cost of $343.5 million.

One New Frontiers mission is launched about every 3 years, such as the New Horizons mission to Pluto.

New Frontiers Program, http://newfrontiers.nasa.gov/

New Horizons, http://pluto.jhuapl.edu/index.php

The Discovery and New Frontiers Programs, Marshall Space Flight Center, http://www.nasa.gov/centers/marshall/moonmars/solar.html

"Cassini shows new angles of Saturn," The Associated Press, March 3, 2007. http://www.nasa.gov/mission_pages/cassini/main/index.html, http://www.tinyurl.com/2uj6m

[166] Iridium, http://www.iridium.com/

Mellow, Craig, "The Rise and Fall and Rise of Iridium," *Air & Space*, September 1, 2004, http://www.airspacemag.com/space-exploration/iridium.html

Barboza, David, "Can Iridium Satellite Service Achieve a Soft Landing?," *The New York Times*, September 7, 1999, http://partners.nytimes.com/library/tech/99/09/biztech/articles/07iridium.html

[167] Globalstar, http://www.globalstar.com/

"Alcatel Alenia Space signs a Euro 661 million contract with Globalstar to build their second-generation LEO satellite constellation," Alcatel-Lucent Press Release, December 4, 2006, http://www.alcatel-lucent.com/wps/portal/!ut/p/kcxml/04_Sj9SPykssy0xPLMnMz0vM0Y_QjzKLd4w3cQ7SL8h2VAQAu32oaA!!?LMSG_CABINET=Docs_and_Resource_Ctr&LMSG_CONTENT_FILE=News_Releases_2006/News_Article_000023

Clark, Stephen, "Soyuz lofts replacement satellites for Globalstar," Spaceflight Now, May 29, 2007, http://spaceflightnow.com/news/n0705/29globalstar/

Lisi, M., and G. Manoni, "High-rate production and testing of spacecrafts and active antennas for mobile/personal satcoms," IEEE Xplore, Proceedings of the Second European Workshop on Mobile/Personal Satcoms, 1996, Pages 203-210, http://ieeexplore.ieee.org/Xplore/login.jsp?url=/iel5/6955/18718/00864062.pdf?tp=&isnumber=&arnumber=864062

[168] Small Satellites, Utah State University, http://www.smallsat.org/

Leonard, David, "The Smallsat Search for Low-Cost Launch," Space News, SPACE.com, August 14, 2006, http://www.spacenews.com/archive/archive06/SmallSatLeonard_0814.html

Stemp-Morlock, Graeme, "Tiny Satellites Promise Low-Risk, Low-Cost Space Future," *National Geographic News*, August 20, 2008, http://news.nationalgeographic.com/news/2008/08/080820-small-satellites.html

Hsu, Jeremy, "Satellite designed to spot asteroid Armageddon," Space.com, July 23, 2008, http://www.msnbc.msn.com/id/25816106/

[169] NASA New Millenium Program, http://nmp.nasa.gov/

Magnuson, Stew, "Teams To Vie For NASA Technology Mission," Spacenews.com, 01 February 2001, http://www.space.com/businesstechnology/technology/new_millennium_teams_010201.html

Carlisle, Candace, "*Space Technology 5* Mission Overview and Lessons Learned," NASA/Goddard Space Flight Center, November 6, 2007, http://multiscstudy.jhuapl.edu/2007_Conf/files/2-ST5_Multi_SC_Conf.ppt

Space Technology 8, NASA, http://nmp.jpl.nasa.gov/st8/index.html

Gai, E., "The century of inertial navigation technology," *Aerospace Conference Proceedings*, 2000 IEEE, Volume 1, Issue 2000, pp. 59-60.

Britt, Robert Roy, "Powering the Future: Soup-Can Spacecraft and Postage-Stamp Engines," SPACE.com, 08 August 2001, http://www.space.com/businesstechnology/technology/jpl_brophy_010808-1.html

Leonard, David, "NSF Taps Tiny CubeSats for Big Space Science," SPACE.com, 27 August 2008, http://www.space.com/businesstechnology/080827-nsf-cubsats-science.html

[170] Dylewski, Adam, "Key advance toward 'micro-spacecraft'," American Chemical Society, August 19, 2008, http://portal.acs.org/portal/acs/corg/content?_nfpb=true&_pageLabel=PP_ARTICLEMAIN&node_id=222&content_id=WPCP_010576&use_sec=true&sec_url_var=region1

Choi, Charles Q., "New Thin Skin to Protect Tiny Spacecraft," SPACE.com, August 19, 2008, http://www.space.com/businesstechnology/080819-acs-micro-spacecraft.html

[171] Christensen, Bill, "Micro Spacecraft To Explore Planets," Technovelgy.com, 10 June 2005, http://www.space.com/businesstechnology/technology/technovel_blackbox_050610.html

[172] Leitner, Jesse, Frank Bauer, David Folta, Russell Carpenter, Mike Moreau, and Jonathan How, "Formation flight in space: Distributed spacecraft systems develop new GPS capabilities," *GPS World*, February 1, 2002, http://www.gpsworld.com/gps/application-challenge/formation-flight-space-727

"NASA, Google Unveil Mars in 3-D," Space.com, February 2, 2009, http://www.space.com/news/090202-nasa-google-mars3d.html

"NASA and Google Launch Virtual Exploration of Mars," Ames Research Center Press Release, February 2, 2009, http://www.spaceref.com/news/viewpr.html?pid=27489

Google Lunar X Prize, http://www.googlelunarxprize.org/

[173] "NASA Tests Interplanetary Internet," Space.com, 19 November 2008, http://www.space.com/news/081119-deep-space-internet.html

Leonard, David, "Red Planet Bound: *Mars Reconnaissance Orbiter*," Space.com, 13 October 2004, http://www.space.com/businesstechnology/technology/mro_tech_041013.html

"Interplanetary Internet Project," Internet Society, http://www.ipnsig.org/ home.htm

Gray, Rich, "On the Edge: Interplanetary Internet," Space.com, 02 May 2003, http://www.space.com/businesstechnology/technology/ontheedge_0305.html

"*Mars Reconnaissance Orbiter*," Jet Propulsion Laboratory, NASA, http://mars.jpl.nasa.gov/mro/

"Lunar Exploration Objectives," NASA, http://www.nasa.gov/pdf/163560main_LunarExplorationObjectives.pdf

[174] David, Leonard, "Extraterrestrial Resources: 'Living Off the Land'," Space.com, 14 November 2003, http://www.space.com/businesstechnology/technology/space_resources_031114.html

[175] "The Tunguska Event—100 Years Later," Science@NASA, June 30, 2008, http://science.nasa.gov/headlines/y2008/30jun_tunguska.htm

Vergano, Dan, "Asteroid anniversary recalls Earth's rocky history," *USAToday.com*, June 30, 2008, http://www.usatoday.com/tech/science/columnist/vergano/2008-06-29-asteroid-anniversary_N.htm

Choi, Charles Q., "Huge Tunguska Explosion Remains Mysterious 100 Years Later," Space.com, June 30. 2008, http://www.space.com/news/080630-mm-tunguska-mystery.html

[176] Kendall, Anthony, "Resource Scarcity and Asteroid Mining," February 5, 2006, http://www.anthonares.net/2006/02/resource-scarcity-and-asteroid- mining.html

Sonter, Mark, "Asteroid Mining: Key to the Space Economy," *ad Astra*, The National Space Society, 09 February 2006, http://www.space.com/adastra/060209_adastra_mining.html

Blair, B.R., "The Role of Near-Earth Asteroids in Long-Term Platinum Supply," Space Resources Roundtable 2, p. 5, Colorado School of Mines, 01/2000, http://adsabs.harvard.edu/abs/2000srrt.conf....5B

White, Bill, "Priming the pump for lunar PGM mining," *The Space Review*, October 24, 2005, http://www.thespacereview.com/article/479/1

Lewis, John S., and Ruth A. Lewis, *Space Resources*, Columbia University Press, New York, 1987.

Lewis, John S., *Mining the Sky: Untold Riches from the Asteroids, Comets, and Planets*, Perseus Books Group, 1997

[177] "NASA Near-Earth Object Survey and Deflection Analysis of Alternatives Report to Congress," NASA Press Release, March 9, 2007, http://www.spaceref.com/news/viewpr.html?pid=22088

The "Safeguard Survey" is currently funded at $4.1 million/year until 2012 to locate problem asteroids.

"2006 Near-Earth Object Survey and Deflection Study," NASA HQ, PA&E, 28 December 2006.

"Asteroid Exploded in Earth's Atmosphere," Space.com, October 8, 2008, http://www.space.com/spacewatch/081008-asteroid-exploded.html For the first time, an asteroid was located before entering the Earth's atmosphere and the time and location of the event predicted. A "kitchen table" sized asteroid exploded over Sudan in Africa on October 7, 2008. It did not impact the Earth.

Reilly, Michael, "Small asteroids can pack a mighty punch," *New Scientist*, December 18, 2007, http://space.newscientist.com/article/mg19626354.400-small-asteroids-can-pack-a-mighty-punch.html

Easterbrook, Greg, "The Sky is Falling," *The Atlantic*, June 2008

Stone, Richard, "Target Earth," *National Geographic*, August 2008, http://ngm.nationalgeographic.com/2008/08/earth-scars/stone-text

"Near-Earth Asteroid Tracking," http://neat.jpl.nasa.gov/

"NEAR-Shoemaker Lands on Asteroid Eros," Space Today, http://www.spacetoday.org/SolSys/Asteroids/NEAR.html

Lauretta, Dante S., "Energy Minerals in Near-Earth Asteroids," *Search and Discovery*, article #80054, August 26, 2009, http://www.searchanddiscovery.com/documents/2009/80054lauretta/ndx_lauretta.pdf

"Discovery is NEAR," http://near.jhuapl.edu/

"Discovery Mission: *Near Earth Asteroid Rendezvous*," http://discovery.nasa.gov/near.html

"Falcon Bringing Home an Asteroid Sample," Space Today online, http://www.spacetoday.org/Japan/Japan/MUSES_C.html

[178] *Armageddon*, http://en.wikipedia.org/wiki/Armageddon_(film)

"NASA Near-Earth Object Survey and Deflection Analysis of Alternatives Report to Congress," March 9, 2007, http://www.nasa.gov/pdf/171331main_NEO_report_march07.pdf

"ESA selects targets for asteroid-deflecting mission *Don Quijote*," esa News, European Space Agency, ESA PR 41-2005, http://www.esa.int/esaCP/SEML9B8X9DE_index_0.html, The plan includes launching two spacecraft at an asteroid—one to measure the orbit and physical characteristics, and a second to impact the asteroid.

[179] Motta, Mary, "Many Pennies From Heaven: Asteroid Impacts Render Riches," Space.com, 17 February 2000, http://www.space.com/businesstechnology/business/asteroid_impact_000216.html

"Automakers think small as precious metal prices soar," Reuters, June 26, 2008, http://www.reuters.com/article/ousiv/idUSSP24374920080626

Kim, Chang-Ran, "Automakers turn to 'nanotechnology' as precious metal prices soar," Reuters, Mineweb, 26 June 2008, http://www.mineweb.com/mineweb/view/mineweb/en/page35?oid=55405&sn=Detail, "The issue of rare metals and rare earth materials is going to be a huge concern for the manufacturing sector," according to Takeshi Uchiyamada of Toyota.

Faughnan, Barbara, and Gregg Maryniak, eds., "Space Manufacturing 7: Space Resources to Improve Life on Earth," Proceedings of the Ninth Princeton/AIAA/SSI Conference, May 10-13, 1989, http://ssi.org/?p=34

[180] Gennery, Donald B., "NASA NEO News: Deflection Scenarios for *Apophis*," SpaceRef.com, August 9, 2005, http://www.spaceref.com/news/viewsr.html?pid=17666

[181] Wakefield, Julie, "Researchers and space enthusiasts see helium-3 as the perfect fuel source," Space.com, 30 June 2000, http://www.space.com/scienceastronomy/helium3_000630.html

Lowman, Paul D., Jr., "Why Go Back to the Moon?," NASA, 14 January 2008, http://www.nasa.gov/centers/goddard/news/series/moon/why_go_back.html

Marburger, John, "Keynote Address," 44th Robert H. Goddard Memorial Symposium, Greenbelt, Maryland, March 15, 2006, http://www.nss.org/resources/library/spacepolicy/marburger1.html

Hsu, Jeremy, "Water Discovered in Moon Samples," Space.com, 09 July 2008, http://www.space.com/scienceastronomy/080709-moon-water.html

[182] Dinerman, Taylor, "Finishing the space station," *The Space Review*, September 17, 2007, http://www.thespacereview.com/ article/956/1

Hurtak, J.J., "Existing Space Law Concepts and Legislation Proposals," The Academy for Future Science, 2005, http://www.affs.org/html/ existing_space_law_concepts.html

"Kennedy Space Center Story," NASA/Kennedy Space Center, 1991, http://www.nasa.gov/centers/kennedy/about/history/story/ch14.html

Kornfeld, Dale M., "Monodisperse Latex Reactor (MLR) A Materials Processing Space Shuttle Mid-Deck Payload," NASA TM-86487, January 1985, NASA, Marshall Space Flight Center. http://ntrs.nasa.gov/archive/nasa/casi.ntrs.nasa.gov/19850012877_1985012877.pdf

"Space Manufacturing," Jim Kingdom, http://www.panix.com/~kingdon/space/manuf.html

[183] Burj Dubai, http://www.burjdubai.com/

"List of world's largest domes," Wikipedia, http://en.wikipedia.org/wiki/List_of_world's_ largest_domes

"Oil Tanker," Wikipedia, http://en.wikipedia.org/wiki/Oil_tanker

McCullagh, Declan, "Next Frontier: 'Seasteading' the Oceans," CNET Tech News, February 2, 2009, http://www.cbsnews.com/stories/2009/02/02/tech/cnettechnews/main4769336.shtml

Freedom Ship International, http://www.freedomship.com/

[184] U.S. Pavilion at Expo '67, http://www.greatbuildings.com/buildings/US_Pavilion_at_Expo_67.html

Buckminster Fuller Dome over Manhattan, *Audubon Magazine*, http://www.audubonmagazine.org/webexclusives/images/bucky-DomeOverManhattan.jpg, http://2.bp.blogspot.com/_fEglkkU__oM/SlT0XfNiv_I/AAAAAAAAF04/TWDvxZTHm5w/s800/Shoji+Sadao+Dome+-+via+NeutralSurface.JPG

[185] Grima, Joseph, "Buckminster Fuller," Icon Eye, http://www.iconeye.com/index.php?view=article&catid=1%3Alatest-news&layout=news&id=3379%3Areview-buckminster-fuller&option=com_content&Itemid=18,

http://stevendejonckheere.blogspot.com/2006/08/cloud-nine.html

http://www.buckminster.info/Ideas/08-IcosDomeCityCloud.htm

Fuller, R. Buckminster, *Critical Path*, St. Martin's Press, 1981, p. 336-337

Baldwin, J., *Bucky Works: Buckminster Fuller's Ideas for Today*, John Wiley & Sons, Inc., 1996, p. 190-191

[186] "How Flying Cars Will Work," howstuffworks.com, http://auto.howstuffworks.com/flying-car.htm

Stephens, Challen, "Cars that fly? It's not pie in the sky," *The Huntsville Times*, July 24, 2005, http://www.macroindustries.com/website/ hsv_times_skyrider_story.htm

Skyrider, Macro Industries, http://www.macroindustries.com/website/files/skyrider/sr-index.htm

[187] O'Neill, Gerard K., *The High Frontier: Human Colonies in Space*, William Morrow & Co, 1976.

Aldrin, Buzz, and David Noland, "Buzz Aldrin's Roadmap To Mars," *Popular Mechanics*, December 2005, http://www.popularmechanics.com/science/air_space/2076326.html

[188] Kolm, Henry, "Mass Driver Up-date," National Space Society, September 1980, http://www.nss.org/settlement/L5news/1980-massdriver.htm

[189] "Solar Cell," http://en.wikipedia.org/wiki/Solar_cell

"New Flexible Plastic Solar Panels Are Inexpensive And Easy To Make," *Science Daily*, July 19, 2007, http://www.sciencedaily.com/releases/2007/07/070719011151.htm

[190] "Space-based solar power," http://en.wikipedia.org/wiki/Solar_power_satellite

Glaser, P.E., "Power from the Sun: Its Future," *Science*, Vol. 162, 957-961, 1968.

Glaser, Peter, E., Frank P. Davidson, and Katinka Csigi, *Solar Power Satellites: A Space Energy System for Earth*, John Wiley & Sons, 1998, http://www.astrobooks.com/index.asp?PageAction=VIEWPROD& ProdID=931

Glaser, Peter E., "The World Needs Energy from Space," Space.com, 23 February 2000, http://www.space.com/opinionscolumns/opinions/glaser_000223.html

Dickinson, R.M,, and C.W. Brown, "Radiated Microwave Power Transmission System Efficiency Measurements," NASA Technical Memorandum 33-727, JPL

"2000 ASTM Standard Extraterrestrial Spectrum Reference E-490-00," http://rredc.nrel.gov/solar/specta/am0/ASTM2000.html

"Space Solar Power," The National Space Society, http://www.nss.org/settlement/ssp/

[191] Mankins, John C., "A Fresh Look at Space Solar Power: New Architectures, Concepts, and Technologies," IAF-97-R.2.03, 38th International Astronautical Federation, http://www.nss.org/settlement/ssp/library/1997-Mankins FreshLookAtSpaceSolarPower.pdf

Gerard K. O'Neil also promoted SBSP, http://ssi.org/

In 2007 the DOD "issued a 75-page study conducted for its National Security Space Office concluding that space power ... offers a potential energy source for global U.S. Military operations. It could be done with today's technology ..." http://www.acq.osd.mil/nsso/solar/SBSPInterimAssesment0.1.pdf

Shiner, Linda, "Where the sun does shine," *Air & Space*, June/July 2008, http://www.airspacemag.com/space-exploration/Sun_Does_Shine.html

"Special Report on Space-Based Solar Power," *Ad Astra*, National Space Society, Spring 2008. http://www.nss.org/adastra/AdAstra-SBSP-2008.pdf

"Space-Based Solar Power Breakthrough to Be Announced," National Space Society, September 9, 2008, http://www.spaceref.com/news/viewpr.html? pid=26383

Nansen, Ralph, *Sun Power: The Global Solution for the Coming Energy Crisis*, Ocean Press, 1995, http://www.nss.org/settlement/ssp/sunpower/index.html

"Solar Power Satellites," European Space Agency, video, http://www.nss.org/settlement/ssp/esavideo.htm

"Space Solar Power: Exploring New Frontiers for Tomorrow's Energy Needs," NASA video, 2002, http://www.nss.org/settlement/ssp/sspnasavideo.htm

The Space Solar Alliance for Future Energy (SSAFE), http://ssafe.wordpress.com/

"Space-Based Solar Power," Space Frontier Foundation, http://spacesolarpower.wordpress.com/

Schirber, Michael, "How Satellites Could Power the Future," LiveScience.com, June 18, 2008, http://www.livescience.com/environment/080618-pf-space-solar.html

Globus, Al, "Solar Power From Space: A Better Strategy for America and the World?," *ad Astra*, National Space Society, May 17, 2007, http://www.space.com/adastra/070517_adastra_solarpowersats.html

Dye, Lee, "NASA Investigates Beaming Energy From Space," ABC News.com, May 16, 2008. http://abcnews.go.com/Technology/story?id=98547&page=1

[192] Hanley, Charles J., "Future may bring us 'space power'," Associated Press, December 28, 2007.

Cho, Dan, "Pentagon backs plan to beam solar power from space," *New Scientist*, 11 October 2007, http://environment.newscientist.com/channel/earth/energy-fuels/dn12774-pentagon-backs-plan-to-beam-solar-power-from-space.html

Cho, Dan, and David Cohen, "Plugging into the Sun," *New Scientist*, 24 November 2007, http://environment.newscientist.com/channel/earth/energy-fuels/mg19626311.600-plugging-into-the-sun.html, PV arrays of several square kilometers placed at geostationary orbit would send the power to Earth as laser light or microwaves for conversion back to electricity.

[193] Cho, Dan, "Pentagon backs plan to beam solar power from space," *New Scientist*, 11 October 2007, http://environment.newscientist.com/channel/earth/energy-fuels/dn12774-pentagon-backs-plan-to-beam-solar-power-from-space.html

"Space Based Solar Power as an Opportunity for Strategic Security," National Security Space Office, 10 October 2007, http://www.nss.org/settlement/ssp/library/nsso.htm

[194] Spires, Shelby G., "NASA braces for a fall," *The Huntsville Times*, February 1, 2010

Borenstein, Seth, and Alicia Chang, "Taxi! Taxi! Will NASA flag a ride into space?," Associated Press, February 1, 2010

[195] Forward, Robert L., and Hans P. Moravec, "Space Elevators," March 22, 1980, http://www.frc.ri.cmu.edu/~hpm/project.archive/1976.skyhook/1982.articles/elevate.800322

Edwards, Bradley C., "The Space Elevator NIAC Phase II Final Report," NASA Institute for Advanced Concepts, March 1, 2003, http://www.niac.usra.edu/files/studies/final_report/521Edwards.pdf, p. 15

[196] Clarke, Arthur C., "The Space Elevator: 'Thought' Experiment or Key to the Universe?," Address to the XXXth International Astronautical Congress, Munich, 20 September 1979, Advances in Earth Oriented Applied Space Technologies, Vol. 1, pp. 39 to 48, Pergamon Press Ltd., 1981, http://www.islandone.org/LEOBiblio/CLARK3.HTM

[197] David, Leonard, "Orbital Debris Cleanup Takes Center Stage," SPACE.com, September 25, 2009, http://www.spacenews.com/civil/orbital-debris-cleanup-takes-center-stage.html

[198] David, Leonard, "Who Owns the Moon?," SPACE.com, December 10, 2008, http://www.space.com/missionlaunches/081210-who-owns-moon.html

Pop, Virgiliu, "Who Owns the Moon?: Extraterrestrial Aspects of Land and Mineral Resources Ownership," Space Regulations Library, Springer-Verlag, New York, LLC, 2008. "Space Stations and the Law: Selected Legal Issues" August 1986, NTIS #PB87-118220

Snead, Mike, "U.N. Law of the Sea Convention and America's spacefaring future," Spacefaring America, June 15, 2007, http://spacefaringamerica.net/2007/06/15/8--un-law-of-the-sea-convention-and-americas-spacefaring-future.aspx

"Oceans and Law of the Sea," United Nations, Division for Ocean Affairs and the Law of the Sea, http://www.un.org/Depts/los/index.htm

"The Antarctic Treaty," National Science Foundation, Office of Polar Programs (OPP), http://www.nsf.gov/od/opp/antarct/anttrty.jsp

Hurtak, J.J., "Existing Space Law Concepts and Legislation Proposals," The Academy for Future Science, 2005, http://www.affs.org/html/existing_space_law_concepts.html

"Treaty on Principles Governing the Activities of States in the Exploration and Use of Outer Space, Including the Moon and Other Celestial Bodies," U.S. Department of State, http://www.state.gov/t/isn/5181.htm

Kiefer, Walter S., "Europa and Titan: Oceans in the Outer Solar System?," Space Science Reference Guide, Second Edition, Lunar and Planetary Institute, 2003, http://www.lpi.usra.edu/science/kiefer/Education/SSRG2-Europa/europa.html

"Is There Life On Jupiter's Moon Europa? Finding Signs Of Current Geological Activity On A Frozen World," *Science Daily*, August 7, 2008, http://www.sciencedaily.com/releases/2008/08/080806210116.htm

Wilson, Edward O., *The Future of Life*, Knopf, 2002.

Arnett, Bill, "Europa," The Nine Planets, 2008, http://www.nineplanets.org/europa.html

[199] Cody, Edward, "China Confirms Firing Missile to Destroy Satellite," Washington Post, January 24, 2007, http://www.washingtonpost.com/wp-dyn/content/article/2007/01/23/AR2007012300114.html

Global Zero, http://www.globalzero.org

"New Push To Eliminate Nukes Worldwide," The Associated Press, CBSNews.com, December 6, 2008, http://www.cbsnews.com/stories/2008/12/06/world/main4652252.shtml?tag=lowerContent;homeSectionBlock202

Brinton, Turner, "Obama's Proposed Space Weapon Ban Draws Mixed Response," *Space News*, Space.com, 4 February 2009, http://www.space.com/news/090204-obama-space-weapons-response.html

Listner, Michael, "A bilateral approach from maritime law to prevent incidents in space," SpaceReview.com, February 16, 2009, http://www.thespacereview.com/article/1309/1

[200] "The Camp David Accords," Jimmy Carter Library,
http://www.jimmycarterlibrary.gov/documents/campdavid/

Talbott, Strobe, Laurence I. Barrett, "Time to START, Says Reagan," *Time*, May 17, 1982, http://www.time.com/time/magazine/article/0,9171,921207,00.html

"United Nations Environment Programme," Ozone Secretariat, http://ozone.unep.org/

"The Montreal Protocol on Substances that Deplete the Ozone Layer," as amended, United Nations Environment Programme, Ozone Secretariat, 2000, http://www.unep.org/OZONE/pdfs/ Montreal-Protocol2000.pdf

[201] Anderson, Greg, "Long-term decisions, short-term politics," *The Space Review*, July 14, 2008, http://www.thespacereview.com/article/1168/1

[202] "NASA Announces Commercial RLV Technology Roadmap Project," NASA Press Release, October 13, 2009, http://www.spaceref.com/news/viewpr.html?pid=29390

Garver, Lori, "Remarks by NASA Deputy Administrator Lori Garver at the 13th Annual FAA AST Space Transportation Conference," SpaceRef.com, February 13, 2010, http://www.spaceref.com/news/viewsr.html?pid=33479

[203] NASA Workforce History, http://wicn.nssc.nasa.gov/cognos/cgi-bin/ppdscgi.exe?
BZ=1AAAABPly7 E~IABEwU6VFChhEnYeaEGVJHTJm5v_6YbQMGmjRz6Ly
Rk8eEkCg9ysCAQSJmH4yYbZAgQbKvZnf9~829TsySWZnFZEWCSpo2Zd9BKSM
nzRsyJWDkCMKG7bshZdzQbWptNq9JkSBTqkgpMvW1U8LZNQNmX5mYfWK7c
mqDmEQDElKvJPN~XbYXBYsWNmPaGHGjZYoQIVpkWCkxMyEMhbGAUxgL7
mH8IaaXGatm~aTdAf~=

[204] Montgomery, Dave, "Aerospace industry fears aging workforce's impact," McClatchy Newspapers, January 20, 2008.
http://www.mcclatchydc.com/staff/dave_montgomery/story/25036.html

Kaplan-Leiserson, Eva, "Mind the Gap," *PE The Magazine for Professional Engineers*, National Society of Professional Engineers, Jan/Feb 2008, p. 30-33

[205] *Rising Above the Gathering Storm: Energizing and Employing America for a Brighter Economic Future*, National Academies Press, 2007,
http://www.nap.edu/catalog.php?record_id=11463#description

Thilmany, Jean, "Catching Them Younger," *Mechanical Engineering*, May 2003, pp. 56-58. Partnership for Innovative Learning sponsored by PTC of Waltham, MA (a software company) provided CAD software to high schools, to introduce students to the process of engineering. "By learning CAD early, students are better prepared for college engineering courses." "Learning to design on a CAD system (also) teaches students to think creatively, to ask questions, and to find answers to their own questions." Having the exposure to engineering in high school better prepares students for the rigors of college engineering courses. "Incorporating more exciting, relevant learning coupled with current technology earlier in a student's life would do wonders for retention of the best students, both male and female."

[206] Freedom Writers Foundation, http://www.freedomwritersfoundation.org

LaGravenese, Richard, "Freedom Banned," *Huffington Post*, July 1, 2008, http://www.huffingtonpost.com/richard-lagravenese/emfreedomem-banned_b_110299.html

[207] Lamb, Gregory M., "How to go to M.I.T. For Free," *The Christian Science Monitor*, January 4, 2007, http://www.csmonitor.com/2007/0104/p13s02-legn.html

"MIT Open Courseware," Massachusetts Institute of Technology, Boston, MA, http://ocw.mit.edu/OcwWeb/web/home/home/index.htm

[208] Grasso, Domenico, Melody Brown Burkins, Joseph Helble, and David Martinelli, "Dispelling the Myths of Holistic Engineering," *PE The Magazine for Professional Engineers*, National Society of Professional Engineers, August/September 2008, p. 26-29. http://www.uvm.edu/-cems/explore/ dispelling.pdf

[209] International Space University, http://www.isunet.edu/

Cockell, Charles S., *Space on Earth: Saving Our World While Seeking Others*, Macmillan Science, 2009, http://www.palgrave.com/products/title.aspx?is=023000752X

[210] Rathbone, Emma, "Almost All Air," *The University of Virginia Magazine*, Winter 2006, http://archives.uvamagazine.org/site/c.esJNK1PIJrH/b.2180635/k.1FAC/9658_Research__Discovery.htm

[211] Stemp-Morlock, Graeme, "CDs and DVDs battle climate change," Cosmos Online, 14 April 2008, http://www.cosmosmagazine.com/news/1938/cds-and-dvds-battle-climate-change

Marshall, Jessica, "Carbon Dioxide: Good for Something?," *Discovery News*, April 10, 2008, http://dsc.discovery.com/news/2008/04/10/carbon-dioxide-plastic.html

Novomer, http://www.novomer.com/

[212] "Rutan Voyager," Smithsonian National Air & Space Museum, http://www.nasm.si.edu/collections/artifact.cfm?id=A19880548000

[213] Michael Coren, "*SpaceShipOne* lands after heart-stopping ride," CNN, October 1, 2004. http://www.cnn.com/2004/TECH/space/09/29/spaceshipone.attempt.cnn/index.html

[214] Malik, Tariq, "Virgin Galactic Unveils Suborbital Spaceliner Design," *Space News*, Vol. 19, Iss.4, p. 17, January 28, 2008.

Virgin Galactic website, "Overview: Who is Involved?," http://www.virgingalactic.com/flash.html

[215] "Dennis Tito," Space.com, http://www.space.com/dennistito/

"Space Adventures Client, Private Astronaut Richard Garriott, Successfully Launches to the *International Space Station*," Press Release, SpaceRef.com, October 12, 2008, http://www.spaceref.com/news/viewpr.html?pid=26675 Richard Garriott, the son of Owen Garriott who was a *Skylab* astronaut in 1974, performed scientific research and commercial experiments during his stay on the *ISS*.

"First Child of U.S. Astronaut Lifts Off," Associated Press, October 12, 2008, http://www.cbsnews.com/stories/2008/10/12/tech/main4516013.shtml

Atkinson, Nancy, "Space Tourist Flights to *ISS* Still On, Says Space Ventures," Universe Today, April 3, 2009, http://www.universetoday.com/2009/04/03/space-tourist-flights-to-iss-still-on-says- space-adventures/

Ansari, Anousheh, and Homer Hickam, *My Dream of Stars*, Palgrave Macmillan, 2010.

[216] "Conestoga," GlobalSecurity.org, http://www.globalsecurity.org/space/systems/conestoga.htm The Conestoga 1620 launch on October 23, 1995 disintegrated in midair 46 seconds after launch. This was the only flight of the Conestoga 1620.

Rocketplane Kistler, orbital flight, http://www.rocketplanekistler.com/

Rocketplane, suborbital flight,   http://www.rocketplane.com/index.html

Starcraft Boosters, Inc., http://www.starbooster.com/

"Rotary Rocket - Summary," Space and Tech, http://www.spaceandtech.com/spacedata/rlvs/rotary_sum.shtml

Leonard, David, "Rotary CEO Quits Amid Rocket Delays," Space.com, June 26, 2000, http://www.space.com/businesstechnology/business/roton_rocket_000626.html

Hynes, Patricia C., Ph.D., "Visionaries of commercial spaceflight," *The Space Review*, September 8, 2008, http://www.thespacereview.com/article/1203/1

[217] Leonard, David, "Exclusive: Rules Set for $50 Million 'America's Space Prize'," Space News, Space.com, November 8, 2004, http://www.space.com/spacenews/businessmonday_bigelow_041108.html

Leonard, David, "Bigelow Aerospace does rocket reality check," Space.com, September 28, 2007, http://www.msnbc.msn.com/id/21039277/

Malik, Tariq, "America's Space Prize: Reaching Higher Than Sub-Orbit," Space.com, October 6, 2004, http://www.space.com/businesstechnology/technology/spaceprize_techwed_041006.html

Malik, Tariq, "New $50 Million Prize for Private Orbiting Spacecraft," Space.com, September 27, 2004, http://www.space.com/missionlaunches/bigelow_spaceprize_040927.html

Boyle, Alan, "Bigelow Shoots for the Moon," Cosmic Log, MSNBC, February 22, 2007, http://cosmiclog.msnbc.msn.com/archive/2007/02/22/65477.aspx

[218] Space X, http://www.spacex.com

Schwartz, John, "Launch of Private Rocket Fails; Three Satellites Were Onboard," *The New York Times*, August 3, 2008, http://www.nytimes.com/2008/08/03/science/space/03launchweb.html?partner=rssyahoo&emc=rss

Klotz, Irene, "Web Entrepreneur Wants NASA to Use His Rockets," *Discovery News*, September 23, 2008, http://dsc.discovery.com/news/2008/09/23/spacex-rocket-nasa.html

[219] Stone Aerospace, http://www.stoneaerospace.com

"Stone Aerospace Announces Formation of a company to Establish a Commercial Refueling Station in Low Earth Orbit (LEO)," Press Release, March 13, 2007, http://www.spaceref.com/news/viewpr.html?pid=22108

[220] "IEEE History Center: Thomas Alva Edison Historic Site at Menlo Park, 1876," Institute of Electrical and Electronics Engineers, http://www.ieee.org/web/aboutus/history_center/menlopark.html

[221] "Northrup Grumman to Debut Earthwatch Educator Program At 2008 National Conference On Aviation and Space Education," press release, October 15, 2008, http://www.spaceref.com/news/viewpr.html?pid=26712

The "Yes I Can..." dolls are a new direction for Mattel, whose 1992 Teen Talk Barbie said several phrases including "Math class is tough." This generated considerable controversy as supporting stereotypes of girls as less capable of learning math and science, discouraging them from pursuing those subjects. The phrase was removed from later dolls.

Jovanovic, Jasna, and Candice Dreves, "Math, Science, and Girls: Can We Close the Gender Gap?," *Connections Newsletter*, National Network for Child Care, 1995, http://www.nncc.org/Curriculum/sac52_math.science.girls.html

Kesner, Kenneth, "Space Center packed with action," *The Huntsville Times*, July 17, 2008, http://www.spacecamp.com/landing/barbie/

[222] "Google initiative would invest millions of dollars in renewable energy," *Daily Designs*, National Society of Professional Engineers, November 28, 2007.

Smith, Rebecca, and Kevin J. Delaney, "Google's Electricity Initiative," *The Wall Street Journal*, November 28, 2007.

Groom, Nichola, "Google looks to develop greener energy sources," Reuters, February 6, 2008, http://www.reuters.com/article/ousivMolt/idUSN0630565920080206

"Google Sponsors Lunar X PRIZE to Create a Space Race for a New Generation," Google Press Release, September 13, 2007, http://www.spaceref.com/news/viewpr.html?pid=23520

Lunar X-Prize Links: http://www.googlelunarxprize.org/, http://www.odysseymoon.com, http://www.spaceref.com/news/viewpr.html?pid=24200

[223] Automotive X Prize, http://www.autoxprize.org/

[224] "What's Your Crazy Green Idea?," X Prize Foundation, http://www.xprize.org/crazy-green-idea

Virgin Galactic, "Overview: Who is Involved?," http://www.virgingalactic.com/flash.html

"Spirit of Innovation Awards," http://www.conradawards.org/

Heinlein Prize, http://www.heinleinprize.com/news/MicrogravityCompetition.pdf

[225] May, Rollo, *The Courage to Create*, W. W. Norton & Company, New York, p. 39, 1975.

[226] "Discovering Linear Perspective," the Renaissance Connection, http://www.renaissanceconnection.org/lesson_art_perspective.html;

Littler, Sarah, "A Linear Perspective to Art," honors paper, Point Loma Nazarene University, San Diego, California, May 15, 2004, http://www2.hmc.edu/www_common/hmnj/littler.pdf

[227] Becker, Barbara J., "Spinning the Web of Ingenuity," History of Technology course lecture University of California, Irvine, Winter Quarter 2004, https://eee.uci.edu/clients/bjbecker/SpinningWeb/lecture10.html

[228] "CXC Biographies: Eileen Collins," http://chandra.harvard.edu/press/ bios/collins_bio.html

[229] "First Native American to Walk in Space," Space Today Online, http://www.spacetoday.org/Astronauts/NativeAmerican.html

"John Herrington's Rocketrek," http://www.rocketrek.com/index.php?pg=rocketrek

[230] "Biographical Data: Peggy Whitson," NASA Johnson Space Center, http://www.jsc.nasa.gov/Bios/htmlbios/whitson.html

[231] Hickam, Homer, *Rocket Boys*, Delacorte Press, 1998.

October Sky, Universal Studios, 1999

[232] Spires, Shelby G., "Cronkite to celebrate space flight," *The Huntsville Times*, December 11, 2007

[233] Schmid, Randolph E., "Early humans almost vanished," Associated Press, April 25, 2008. http://eventhorizon1984.typepad.com/event_horizon_1984_blog/history/

"The Genographic Project," *National Geographic*, https://www3.nationalgeographic.com/genographic/

Stix, Gary, "Traces of a Distant Past," *Scientific American*, July 2008.

"DNA Study Supports African Origin Theory," CBS News, February 22, 2008, http://www.cbsnews.com/stories/2008/02/22/tech/main/3862267.shtml

[234] McKenzie, Richard B., *The Paradox of Progress*, Oxford University Press, 1997.

Margolis, Mac, "Brazil Pays the Poor," Newsweek, December 7, 2009, p. 15.

[235] Address by OSTP Director John Marburger at the Goddard Memorial Symposium, March 7, 2008, http://www.spaceref.com/news/viewsr.html?pid=27253

[236] Prantzos, Nikos, *Our Cosmic Future: Humanity's Fate in the Universe*, Cambridge University Press, 2000 (French edition, Editions du Seuil, 1998), http://assets.cambridge.org/97805217/70989/sample/9780521770989wsc00.pdf

Hsu, Jeremy, "Spaceflight at warp speed? Make it so," Space.com, August 13, 2008, http://www.msnbc.msn.com/id/26179686/

Obousy, Richard K., and Gerald Cleaver, "Warp Drive: A New Approach," December 16, 2007, http://www.scribd.com/doc/1251197/Warp-Drive-A-New-Approach?query2=gerard%20cleaver%20richard%20obousy

[237] Diamond, Jared, *Collapse: How Societies Choose to Fail or Succeed*, Viking Adult, 2004.

Diamond, Jared, *Collapse: How Societies Choose to Fail or Succeed*, Voices, Arts & Lectures, University of California, Santa Barbara, February 17, 2005, http://codesmithy.wordpress.com/2008/08/20/jared-diamond-collapse-lecture/

[238] "*Hubble's* Deepest View Ever of the Universe Unveils Earliest Galaxies," HubbleSite, News Release Number: STSci-2004-07, March 9, 2004, http://hubblesite.org/newscenter/archive/releases/2004/07/text/

"*Hubble* directly observes planet orbiting Fomalhaut," Hubble Information Center, 13 November 2008, http://www.spacetelescope.org/news/html/heic0821.html

"Huge Exoplanet News Items: Pictures!!!," *Discover*, Blogs/Bad Astronomy, http://blogs.discovermagazine.com/badastronomy/2008/11/13/huge-exoplanet-news-items-pictures/

Index

1,000 spacecraft proposal, 135, 164
2001: A Space Odyssey, 14, 24

Abbey, George, 122
aerogels, 78, 173, 174
Aerogels, 173
Airship-To-Orbit, 87
Albert, Wilhelm, 67
Aldrin, Edwin "Buzz", 11, 12, 147
American Institute of Aeronautics and Astronautics, 17
American West, 24, 25, 26, 27, 42, 82, 84, 121, 155, 165, 184, 190, 191, 194
Anderson, Greg, 158
Ansari, Anousheh, 176
Apollo 11, 11, 14, 23, 24, 58, 123, 128, 168
Armstrong, Neil, 1, 23
Arthur, Chester A., 41
Artsutanov, Yuri N., 91
asteroid, 81, 97, 100, 101, 117, 129, 131, 136, 138, 139, 140, 147, 164
  Apophis, 117, 140, 141
  Near Earth Asteroid Rendezvous mission, 139
  Sudbury, 138, 139
  Tunguska, 138
Atkinson, Ken, 90

barnstorming, 32
Bass, Edward, 62
Bigelow, Robert, 176, 177
Bios-3, 62
Biosphere 2, 62-64
Bluford, Guion Jr., 18
Boughman, Ray, 90
Brandenburger, Adam, 115
Branson, Richard, 117, 175, 180
Braungart, Michael, 55
Brazil, 129, 154, 187
Brooklyn Bridge, 67, 93
Brown, Lester, 8, 55
Brunelleschi, Filippo, 181, 182
Burj Khalifa (Dubai), 143
Burkins, Melody, 171
Bush, George H. W., 58, 128
butterfly effects, 64, 65

carbon nanotubes, 78, 89, 90, 92, 94, 96, 173, 174
carbon neutral, 6
Carson, Rachel, 4
Cayley, Sir George, 29
Challenger (Space Shuttle), 18, 19, 44, 45, 47, 53, 56, 58, 79, 85, 107, 112, 170
Challenger Center for Space Science Education, 47
Chandrasekhar, Prasanna, 133
Chanute, Octave, 29

China, 2, 117, 118, 129,
   143, 154, 157, 187
Civil War, U.S., 24, 27
Clarke, Arthur C., 79, 80,
   81, 91, 92, 98, 116, 155,
   170, 193
Coates, Geoffrey, 174
Cockell, Charles, 173
Cold War, 1, 14, 18, 40, 80
Collins, Eileen, 182
Columbia (space shuttle),
   17, 50, 84, 107, 170, 183
Columbia University, 64
commercial space
   Alenia Spazio, 132
   Bigelow Aerospace, 177
   Falcon, 85
   Falcon 1, 176
   Falcon 9, 177
   Genesis I and Genesis II,
      176
   GlobalStar, 131, 132
   Iridium, 131, 132
   latex spheres, 20, 143
   mass production, 131
   miniaturization, 132, 133,
      164
   Orbital Sciences
      Corporation, 85
   Personal Spaceflight
      Federation, 117
   pharmaceuticals, 143
   Shackleton Energy
      Company, 177
   SpaceShipOne, 85, 95,
      117, 175, 176
   SpaceShipThree, 176
   SpaceShipTwo, 175
   SpaceX, 85, 176

   Stone Aerospace, 177
   Sundancer, 176
   tourists, 175
   Virgin Galactic, 117
competitions (see also
   prizes)
   Clarke-Bradbury
      International Science
      Fiction competition, 92
   NASA Centennial
      Challenges, 76, 85, 92
condensation trails
   (contrails), 86
controlled ecological life
   support system (CELSS),
   62
Co-opetition, 112, 115
   game theory, 115, 116
Cronkite, Walter, 13, 184
Crump, William, 63
Culbertson, Frank, 188
Curl, Robert, 89
Curtiss, Glenn H., 31

Darwin, Charles, 174, 188,
   189
daVinci, Leonardo, 28
de Soto, Hernando, 188
Dezhurov, Vladimir, 188
Diamandis, Peter, 179
Diamond, Jared, 191, 192
dust on the Moon, 109

Edison, Thomas, 41, 42, 68,
   69, 75, 168, 177
energy efficiency
   Energy Star, 5, 73
   net-zero energy
      buildings, 73

energy sources
 "flying windmills", 74
 cold fusion, 74
 hydrogen-producing
  algae, 74
 methane hydrates, 73
 osmotic power, 74
 radiation or low-level
  heat, 74
 sea turbines, 74
Environmental Control and
 Life Support System
 (ECLSS). See also Life
 Support
Eratosthenes, 127, 182
Eremets, Mikhail, 87
Evans, Oliver, 48
Exploration Technologies
 program, 61

Federal Aviation
 Administration, 117
Fritts, Charles, 149
Fuller, Buckminster, 4, 74,
 89, 143, 144
 Cloud Nine, 144, 145,
  146, 164
 geodesic, 89, 143, 144,
  145

Gargarin, Yuri, 39
Garriott, Richard, 176
Gehman, Admiral Hal, 50,
 123
Glaser, Peter, 150
Global Energy Network
 Institute (GENI), 7
G,N, & C, 38, 133

Goddard, Robert, 35, 36,
 48, 132, 169
Google, 130, 134, 179
Google Lunar X Prize, 134
Gore, Al, 7
Grasso, Domenica, 171
Greensburg, Kansas, 6-7
Griffin, Michael, 80, 108,
 120, 121
Group of Eight, 54
Gruwell, Erin, 171

Hartmann, William K., 79
Heinlein, Robert, 84, 180
Helble, Joseph, 171
Herrington, John, 182, 183
Hickam, Homer, 184
Hubbert, M. King, 4
Humphries, Randy, 52

India, 2, 118, 129, 154, 187
Institute for Advanced
 Studies in Life Support,
 63
INTELSAT, 116, 158, 164
Intergovernmental Panel on
 Climate Change (IPCC),
 4
Internal Thermal Control
 System (ITCS), 103, 104,
 105, 106, 107
international agreements
 1979 Israel-Egypt Peace
  Treaty, 158
 Copenhagen Climate
  Change Summit, 5
 Kyoto Protocol, 5
 Law of the Sea Treaty,
  157

Montreal Protocol, 158, 165
Oregon Treaty of 1846, 25
Outer Space Treaty, 157
Strategic Arms Limitation Treaty (SALT), 15
Strategic Arms Reduction Talks (START), 18, 158
UN Resolution 1721, 116
weapons in space, 157
International Conference of Life Support and Biosphere Science, 63
International Council of Scientific Unions, 127
International Energy Agency (IEA), 54
International Geophysical Year (IGY), 38, 127, 128, 129
International Partnership for Energy Efficiency Cooperation (IPEEC), 8
International Space Decade, 129, 164
International Space Exploration Coordination Group (ISECG), 118, 119, 156, 157
International Space Station (ISS), 80, 102, 103, 104, 105, 106, 107, 112, 113, 114, 118, 143, 147, 153, 157, 176, 178, 184, 189, 190
biocide, 113

Destiny, 102, 103, 104
International Space University (ISU), 172, 173
International Space Year (ISY), 128
Interplanetary Internet, 135
Irwin, James B., 12, 95
Isaacs, John, 91

Japan, 2, 54, 68, 89, 118, 129, 139, 150, 154, 155
Jefferson, Thomas, 24, 121
jet engine, 33, 84, 123
Jevons Paradox, 76
Johnson Space Center (Johnson), 17
Junior Engineering Technical Society (JETS), 16

Kennedy, John Fitzgerald, 14, 23, 39, 104, 116, 126, 156, 160
Kepler, Johannes, 35, 98
Kevlar, 89, 90, 92, 94, 175
King, Martin Luther, Jr., 4, 14, 203
Kistler, Sam, 173
Korolev, Sergei, 39
Kranzberg, Melvin
First Law of Technology, 77
Kroto, Harold, 89

Lang, Fritz, 37, 48
Langley, Samuel P., 29, 31
Laporte, Amaury, 66

legislation
   Air Commerce Act, 207
   Civil Aeronautics Act,
     207
   Clean Air Act, 15
   Clean Water Act, 15
   Federal Aviation Act, 207
   Homestead Act, 26
   Mining Act, 26
   NASA Authorization Act
     of 2005, 138
   National Environmental
     Policy Act, 14
   Pacific Railway Act, 24,
     26
Leonov, Alexei, 40
Levine, Arnold, 33
Lewis & Clark, 23, 24, 121
life support, vi, 10, 51, 52,
   53, 59, 60, 61, 62, 72,
   108, 110, 113, 114, 168,
   191
lighting
   compact fluorescent, 68
   ideal light bulb, 68
   incandescent, 68
   light emitting diodes
     (LED), 5, 68, 69
   solid-state lighting (SSL),
     68
Lilienthal, Otto, 29, 30, 170
Lincoln, Abraham, 24, 26
Lindbergh, Charles, 32, 34,
   190
Linkin, Vechaslav, 117
Listner, Michael, 158
Livingston, David, 2
Lovins, Amory, 8
Lucid, Shannon, 113

Magellan, Ferdinand, 188
Makhijani, Arjun, 8
Marburger, John, 57, 189
Mars Cyclers, 147
Marshall Space Flight
   Center (Marshall), iv, vi,
   17, 19, 20, 39, 51, 52, 53,
   59, 72, 104, 113, 114
Martinelli, David, 171
Masdar City, 6, 145
mass production, 46, 132,
   133, 153
Mattel, Inc., 179
May, Rollo, 180, 181
McAuliffe, Christa, 45, 47,
   181
McDonough, William, 55
McKibben, Bill, 54
Means, James, 29
Melvill, Mike, 176
Montgomery, Dave, 169
Mormons, 26
motivations
   greed, fear, and curiosity,
     81
   opportunity, 82
Musk, Elon, 176

Nalebuff, Barry, 115
National Advisory
   Committee on
   Aeronautics (NACA), 31,
   32, 33, 39, 165
National Aeronautics and
   Space Administration
   (NASA), 3, 13-15, 17-21,
   23, 24, 39, 40, 52, 53, 56,
   58-63, 76, 80, 85, 92, 93,

98, 100, 108-110, 112-
114, 118, 120, 122, 129,
130, 132, 139-141, 149,
163, 165, 168, 169, 177,
178, 183, 184, 190
NERVA, 99
Newton, Isaac, 33, 35, 36,
183
Nordhaus, Ted, 55
Norris, Pamela, 173

O'Neill, Gerard K., 146
Obama, Barack, 153, 158,
163
Oberth, Hermann, 35, 37,
169
Olsen, Gregory, 176
OpenCourseWare, 171
orbital debris, 93, 136, 155,
157

Pacala, Stephen W., 8
Pearson, Jerome, 91
Perminov, Anatoly, 117
Pickens, T. Boone, 7, 8
Pike, Zebulon, 25
Planetary Society, 98, 117
Poulin, Philippe, 90
power plant
    coal-fired, 69
    geothermal, 71
    hydroelectric, 69
    ideal power plant, 69
    ITER, 71
    natural gas, 69
    nuclear, 69
    nuclear fusion, 71
    photovoltaic (PV), 70,
        149, 150, 151, 164

reduced consumption, 73
solar, 70
wind, 6, 7, 8, 30, 32, 70,
    71, 179
Precision Combustion, Inc.,
    60
prizes
    America's Space Prize,
        176
    Ansari X Prize, 117, 175,
        179
    Automotive X Prize, 179
    Crazy Green Idea, 179
    Energy and Environment
        X Prize, 179
    Google Lunar X Prize,
        178
    Heinlein Prize, 179
    Pete Condad Spirit of
        Innovation Award, 179
    Virgin Group Prize, 179

radiation, 108
    galactic cosmic rays
        (GCR), 108
    solar particle events
        (SPE), 108
Rayman, Marc, 97
Reagan, Ronald, 17, 20, 45,
    56, 57, 84, 128, 142
Reed, Kevin, 150
research
    Defense Advanced
        Research Projects
        Agency (DARPA), 85
Ride, Sally, 18
risk-reduction, 164

risks, 3, 25, 26, 43, 44, 45,
  100, 152, 158, 162, 163,
  165, 186
rocket fuel
  "polymeric" nitrogen, 87
  hydrogen, 87
Roebling, John, 67
Rogers, Tom, 122
Rutan, Burt, 85, 95, 117,
  175, 176
Rutan, Dick, 174

Sakakura, Toshiyasu, 174
Schweikert, Rusty, 140
Scobee, June, 47
selections, 72
  "advantage/disadvantage"
    method, 72
  "weighted factors"
    method, 72
Shellenberger, Michael, 55
Shepard, Alan, 175
Shuttleworth, Mark, 176
Simonyi, Charles, 176
Singapore, 117, 154, 155
Smalley, Richard, 89, 90
Socolow, Robert, 8
Solar System Positioning
  System, 135
Solid Rocket Booster
  (SRB), 45, 46, 47
Southern Exposition, 41
Space Based solar power
  (SBSP), 150
space elevator, 30, 91, 92,
  93, 94, 95, 96, 97, 144,
  145, 147, 153, 155, 164
  Skyhook, 95

Space Exploration Initiative
  (SEI), 56, 58, 59, 63, 112
space missions
  Apollo 8, 15
  Apollo 11, 11, 14, 23, 24,
    58, 123, 128, 168
  Apollo 13, 37
  Apollo-Soyuz Test
    Project, 15, 40
  Cosmos-1, 98
  Dawn, 97
  Deep Space 1 (DS1), 85,
    97
  Discovery program, 130
  Don Quijote, 139
  Explorer 1, 38, 39, 127,
    129
  Hubble Space Telescope,
    18, 20, 56, 79, 81, 193
  Intelsat 1, 116
  Laika (Sputnik II), 38
  Mars Reconnaissance
    Orbiter (MRO), 134,
    135
  MESSENGER, 98
  Mir, 113
  Mir 2, 113
  Near Earth Object
    Surveillance Satellite
    (NEOSSat), 132
  New Frontiers program,
    130
  New Millennium
    program, 132
  Odyssey, 100
  Phobos-Grunt, 117
  Project Prometheus, 100
  Skylab, 15, 16, 17, 52, 53,
    62, 106, 107, 113, 190

*Solar-Terrestrial
    Relations Orbiters
    (STEREO)*, 109
*Space Technology Five
    (ST5)*, 132
Spacelab, 18, 19, 56, 62,
    119, 143
Spacelab 3, 19
Sputnik I, 37, 38, 39
*Tethered Satellite System
    (TSS)*, 94
Vega 1 and *2*, 117
Voyager 1, 18, 98
spacecraft propulsion
    electrodynamic, 95
    ion drive, 97
    methane, 99
    nuclear, 99
    solar sail, 98
Spaceward Foundation, 92
Stafford, Thomas, 40
Stone, Bill, 177

Tai, Alex, 117
Tito, Dennis, 176
Transcontinental Express,
    25
Transcontinental Railroad,
    24, 84
Tsiolkovsky, Konstantin,
    35, 36, 37, 39, 91, 124,
    139, 153, 169
Tunguska, 138
Tyurin, Mikhail, 188

U.S. Department of Energy
    (DOE), 5, 7, 73, 163
U.S. Green Building
    Council

Leadership in Energy and
    Environmental Design
    (LEED), 6, 7
U.S. Green Building
    Council (USGBC), 6
UN Millennium
    Development Goals, 56
Union of Soviet Socialist
    Republics (USSR), 14,
    15, 18, 23, 37, 38, 39, 40,
    62, 75, 79, 80, 81, 107,
    112, 113, 116, 127, 158,
    168

Verne, Jules, 35, 170
von Braun, Wernher, 24, 37,
    38, 39, 169

Walker, Charles, 143
Whitson, Peggy, 183
Wilson, Woodrow, 32
World Trade Center, 1
Wright brothers, 28, 29, 30,
    31, 32, 34, 95, 121, 122,
    169, 170, 175, 189

Yakobson, Boris, 90
Yeager, Jeanna, 174

Zhigang, Zhai, 129